W9-ABM-842

DECOMPOSABILITY

QUEUEING AND COMPUTER SYSTEM APPLICATIONS

ACM MONOGRAPH SERIES

Published under the auspices of the Association for Computing Machinery Inc.

Editor ROBERT L. ASHENHURST *The University of Chicago*

A. FINERMAN (Ed.) University Education in Computing Science, 1968

A. GINZBURG Algebraic Theory of Automata, 1968

E. F. CODD Cellular Automata, 1968

G. ERNST AND A. NEWELL GPS: A Case Study in Generality and Problem Solving, 1969

M. A. GAVRILOV AND A. D. ZAKREVSKII (Eds.) LYaPAS: A Programming Language for Logic and Coding Algorithms, 1969

THEODOR D. STERLING, EDGAR A. BERING, JR., SEYMOUR V. POLLACK, AND HERBERT VAUGHAN, JR. (Eds.) Visual Prosthesis: The Interdisciplinary Dialogue, 1971

JOHN R. RICE (Ed.) Mathematical Software, 1971

ELLIOTT I. ORGANICK Computer System Organization: The B5700/B6700 Series, 1973

NEIL D. JONES Computability Theory: An Introduction, 1973

ARTO SALOMAA Formal Languages, 1973

HARVEY ABRAMSON Theory and Application of a Bottom-Up Syntax-Directed Translator, 1973

GLEN G. LANGDON, JR. Logic Design: A Review of Theory and Practice, 1974

MONROE NEWBORN Computer Chess, 1975

ASHOK K. AGRAWALA AND TOMLINSON G. RAUSCHER Foundation of Microprogramming: Architecture, Software, and Applications, 1975

P. J. COURTOIS Decomposability: Queueing and Computer System Applications, 1977

In preparation

ANITA K. JONES (Ed.) Perspectives on Computer Science: From the 10th Anniversary Symposium at the Computer Science Department, Carnegie-Mellon University

JOHN R. METZNER AND BRUCE H. BARNES Decision Table Languages and Systems

Previously published and available from The Macmillan Company, New York City

V. KRYLOV Approximate Calculation of Integrals (Translated by A. H. Stroud), 1962

DECOMPOSABILITY

Queueing and Computer System Applications

P. J. Courtois

MBLE RESEARCH LABORATORY
BRUSSELS, BELGIUM

ACADEMIC PRESS New York San Francisco London 1977

A Subsidiary of Harcourt Brace Jovanovich, Publishers

ACADEMIC PRESS, INC.
111 Fifth Avenue, New York, New York 10003

United Kingdom Edition published by
ACADEMIC PRESS, INC. (LONDON) LTD.
24/28 Oval Road, London NW1

Library of Congress Cataloging in Publication Data

Courtois, P J
Decomposability.

(ACM monograph series)
Bibliography: p.
Includes index.
1. Decomposition method. 2. Electronic digital computers–Evaluation. 3. Queuing theory. I. Title. II. Series: Association for Computing Machinery. ACM monograph series.
QA402.2.C68 519.8′2 76-19484
ISBN 0–12–193750–X

PRINTED IN THE UNITED STATES OF AMERICA

Contents

Foreword

Dr. Courtois has produced an important book on the use of computers to study computers. His book is about computation: analytic and numerical methods for examining the behavior of complex systems. At the same time it is about computers: its area of application is the design of computers and the analysis of their performance.

This kind of incestuous or introspective relation of computers to their own kind is not at all unusual—and for a very good reason. Much scientific activity today, in almost all fields of inquiry, is directed toward understanding complexity—the complexity of thunderstorms and ecological systems, the complexity of genetic control of developing organisms, the complexity of human thought processes, and the complexity of computers. Since the study of complex systems calls for new experimental methods and new computational tools, a substantial part of the effort directed toward the study of complexity has been aimed at the development of such tools. In the past twenty-five years, we have seen a great flowering of computational mathematics, providing us with linear, integer, and dynamic programming, with queueing theory, with simulation techniques, and with great advances in the classical methods of numerical analysis.

The computer has played a central role in spurring these developments, for computation with an electronic computer is a very different matter from computation with an old-fashioned desk calculator, calling for new approaches and new points of view, and creating new aspirations as well. The time is long past, of course, since anyone has imagined that computer brute force could substitute for skillful analysis, and thoughtful analysts never held the illusion that such a substitution was possible. Nature has no difficulty whatsoever in inventing systems whose exact analysis defies the most powerful efforts of present or prospective computers—and large computers themselves constitute one class of such systems.

What we cannot do by brute force we must do with cunning. One form of cunning is to recognize that the systems produced by Nature and by the art of man are seldom as complex as they might be, or as their size and number of components might lead us to expect them to be. Fortunately for our prospects of understanding complex systems, it is not common for each element in such systems to interact strongly with each other element. On the contrary, the world, even the world of complexity, is mostly empty. The matrices that describe complex systems tend to be sparse matrices, and very unrandom ones. We can augment our analytic power greatly by exploiting this sparsity and this order.

Dr. Courtois' analysis is based on his important observation that large computing systems can usefully be regarded as nearly completely decomposable systems—systems arranged in a hierarchy of components and subcomponents, with interactions within components that are strong and fast compared with the interactions between components at the same level. Near-decomposability is not peculiar to computing systems: it has been observed in economic structures and in a variety of biological models, genetic and developmental. Only quite recently have its computational implications begun to be explored.

By exploiting with great originality the near-complete decomposability of large computer structures, Dr. Courtois has greatly advanced our ability to compute their behavior, and hence to design them. He has made an important conceptual contribution which is, at the same time, an important practical one. Both hardware and software designers will find in this book a set of powerful methods for systems analysis, as well as a clear exposition of the theory of nearly completely decomposable systems upon which these specific analytic methods are based. While the book addresses itself specifically to computer applications, it should also have substantial interest for scientists investigating other kinds of complex systems, who may find these or similar techniques applicable to the solution of their own computational problems.

Carnegie-Mellon University HERBERT A. SIMON
July 1976

Preface

This monograph groups the results of research work started seven years ago. The intention was to progress toward analysis methods that would be more effective in the evaluation of computer system performance, and which should ultimately lead to the design of optimized systems.

Due to the ever-increasing complexity of computer systems and of their applications, this research found itself quickly and naturally oriented toward methods proceeding by decomposition and approximations; the separation of problems and the conceding of approximations have always been golden rules for tackling the complex reality around us. We first started rather empirically using successive approximations that exploited the great differences of magnitude of the speeds at which information flows at the various levels of a computer memory hierarchy (Courtois and Georges, 1970). But it was the discovery of the aggregation methods widely used in economics and of the results obtained by H. A. Simon and A. Ando (1961) on nearly completely decomposable structures of linear models that made our work take a very decisive turn. It permitted us to justify more rigorously assumptions that we had previously envisaged but empirically, and it enlarged our investigation ground considerably.

The book may be divided into three parts. The first part, being comprised of the first three chapters, is of rather general interest. It aims at gathering together some basic elements of a theory of nearly completely decomposable stochastic matrices. The Simon–Ando theorems are presented, some of their aspects being dealt with in precise detail. Elements of Wilkinson's (1965) perturbation theory inspired us to undertake also a study of the sharpness of the accuracy of the Simon–Ando approximation. This study leads to the definition of criteria of near-complete decomposability.

The second part, which is composed of the next three chapters, is devoted to the analysis of stochastic queueing networks which appear as a type of key model in our work. On the one hand, these models admit, at least in their simplest forms, a matrix representation that allows the theory of decomposability to be exploited straightforwardly; on the other hand, congestion problems in information processing systems may, in many respects, be adequately studied by means of queueing networks models. A method of analysis by decomposition and aggregation for these models is proposed and discussed at length.

The last part of the book deals concretely with the problem of computer system performance evaluation. The material of the first two parts is applied to the analysis of different aspects, hardware and software, of the dynamic behavior of computer systems and user programs. Chapter VIII is a shortened version of a study of program paging behavior made in collaboration with H. Vantilborgh (Courtois and Vantilborgh, 1976); Chapter IX gives a detailed analysis of an aggregative model of a typical multiprogramming time-sharing computing system. It is hoped this last part should illustrate that aggregation is not only an efficient technique for obtaining quantitative results but also for gaining insight and conceptual clarity on the parts played by the many parameters of a complex model.

Finally, a last chapter examines the striking affinity that appears to exist between the concept of aggregate in nearly completely decomposable structures and the notions of module and level of abstraction so frequently invoked in computer system design and software engineering. Variable aggregation appears in this context as a technique that could enable the level by level evaluation of a system to proceed in pace with its level by level design.

In the main, the chief result of this work is that it proposes a rather general approach to computer system model building and suggests a framework in which it seems possible to coordinate future analyses carried out in various directions and at different levels of detail.

Acknowledgments

This book would not have been written without the support of the MBLE Research Laboratory, Brussels, and of its director, Professor V. Belevitch, to whom go my foremost thanks. I should like also to thank D. L. Parnas, who contributed many discussions to the material of Chapter X. Moreover, his invitation to the Computer Science Department of Carnegie–Mellon University during the winter 1971–1972 led to personal contacts with Professor H. A. Simon to whom I owe a great debt of thanks. His advice and encouragement were invaluable for the work (Courtois, 1972) done at Carnegie–Mellon which was already the substance of this monograph.

I am also greatly indebted to my friends of the MBLE Research Laboratory for their help. The work benefited from the early assistance of J. Georges, while H. Vantilborgh followed the manuscript through several changes and is at the origin of numerous improvements; both of them, together with M.-J. Van Camp prepared the computer programs which yielded the numerical data exploited in Chapters XIII and IX, and in Appendix III. In addition I would like to thank Professors J. Meinguet, G. de Ghellinck, G. Louchard, and E. Milgrom who served on the committee for my doctoral dissertation at the Catholic University of Louvain, Belgium, for the useful advice and comments they gave on this occasion.

Finally, my thanks are due to Mrs. A. Toubeau for her patient and expert typing of the manuscript, and to C. Semaille who realized the figures.

Introduction and Overview

> The concept of hierarchic order occupies a central place in this book, and lest the reader should think that I am riding a private hobby-horse, let me reassure him that this concept has a long and respectable ancestry. So much so, that defenders of orthodoxy are inclined to dismiss it as "old hat"—and often in the same breath to deny its validity. Yet I hope to show as we go along that this old hat, handled with some affection, can produce lively rabbits.[1]
>
> A. KOESTLER (1967)
> *The Ghost in the Machine*

The research work reported in the following pages was inspired by the desire of finding adequate methods to evaluate and predict the performances of current general-purpose computer systems.

Fundamentally this problem amounts to constructing appropriate models of the dynamic behavior of these systems. By "model" we mean a set of relations between unknowns (typically, measures of efficiency, such as the system response time or the resource utilization factors) and various parameters that represent the relevant characteristics of the system and of its work load. These relations must be sufficiently simple so as to permit the evaluation of the unknowns; they must at the same time be faithful to the system, i.e., they must capture the relevant laws that govern its mode of operation. Rather obviously, such models with useful and accurate predictive properties cannot be constructed if the system is not first properly understood. To solve the problem of computer performances prediction requires not only powerful evaluation techniques but also a better understanding of the basic principles

[1] Reprinted with the permission of A. D. Peters & Co. Ltd., London.

of system behavior. These two types of requirements define the main lines of conduct of our work.

Analyses of computer system behavior have abounded during the past decade; surveys by Lucas (1971) and Graham (1973), among others, give evidence of this efflorescence. Many of these analyses, however, focus their attention on a particular aspect of system behavior; they analyze a processor-sharing mechanism, a store management strategy, a class of page replacement algorithms, an access method to rotating storage devices, a job scheduling policy, etc. Such investigations are usually carried out in depth, by means of well-defined but simplified models in the context of which appropriate mathematical disciplines can be exploited. They provide a body of knowledge that, to a certain extent, helps to understand how each of these mechanisms operates and compares in efficiency with competitors. These analytical studies made in isolation are, in a way, a normal step in engineering; "scientific progress has been made by analysis and artificial isolation . . .," as Russell (1948) observed.

Nevertheless, these isolated analyses give practically no information on the way the various mechanisms cooperate and interact upon each other within the wholeness of a single system. The influence of a given component upon the behavior of a complete installation, the contribution of a part to the efficiency of the whole are beyond the scope of these investigations. It would be only a lesser evil if computing systems enjoyed what von Bertalanffy (1968) calls the "summativity" character; then, as "a heap of heaps" or a parallelogram of mechanical forces, the variations in the behavior of the total system would merely be the sum of the variations of its elements considered in isolation. But the behavior of a computing system cannot be summed up from its isolated parts; the behavior of the component parts can be quite different within the system from what it is in isolation. The high degree of resource sharing, multiprocessing, and exploitation schemes such as multiprogramming and multiaccess have introduced complex dependencies between the various processes that take place in a modern computing system. In the face of this complexity, the isolated analyses mentioned above leave the designer without means to assess the global consequences of his particular design options. As Simon (1969) put it:

> Only fragments of theory are available to guide the design of a time-sharing system or to predict how a system of a specified design will actually behave in an environment of users who place their several demands upon it. Most actual designs have turned out initially to exhibit serious deficiencies; and most predictions of performance have been startlingly inaccurate.
>
> Under these circumstances, the main route open to the development and improvement of time-sharing systems is to build them and see how they behave. And this is what has been done. They have been built, modified, and improved in successive stages. Perhaps theory could have anticipated these experiments and made them unnecessary. In fact, it didn't; and I don't know anyone intimately acquainted with these exceedingly

complex systems who has very specific ideas as to how it might have done so. To understand them, the systems had to be constructed, and their behavior observed.

Comprehensive analyses of complete systems are of course being attempted to alleviate this state of affairs. These attempts in general resort to simulation techniques. While analytic modeling is limited by the lack of algebraic solutions for complex models, simulation is in principle apt to investigate systems of arbitrary complexity. This generality is, alas, achieved at some expense. The results obtained by simulation are often difficult to interpret with a reasonable degree of confidence. In many cases, above a certain level of complexity, the validity of a simulation model can only be asserted by checking that the model is an exact copy of the real system; as a consequence of this, the analyst is inclined to define and build his model at a level of detail that tends to approach the complexity of the real system. Such detailed models rapidly become difficult to understand and give per se less and less insight into the real system behavior. They yield results that are received with much skepticism; moreover, they quickly become difficult to modify and expensive to adjust to different design alternatives.

The moral is that existing modeling approaches have their own advantages and limitations; and that in face of the different structures of the components of a computing system, and especially of the many different levels of details at which performance prediction needs to take place, no single modeling technique simply asserts itself as being always the most useful. On the contrary, these techniques appear to complement each other.

There are two possible attitudes to take if we want to improve upon this situation: the search for a new evaluation technique more powerful and more general than existing ones, or the search for a general framework in which the different tools of analysis that are already available could be integrated. We opted for this second and less ambitious approach which appeared more promising.

From what has been said, one can already have an inkling of what this second approach may require. We need:

(1) *Criteria* according to which a computing system could be dissected into constituents that can be understood, analyzed, and calibrated separately. Once this is done, our task will be simplified because (a) the model of a constituent is likely to be simpler than the model of the whole, (b) we need not use the same technique of analysis for all constituents, and (c) elaborate and reliable models already exist for certain constituents.

(2) *A model of "macrorelations"* among these constituents so that the results of the isolated analyses can be combined to give an evaluation of the whole system behavior. This macromodel is likely to be a nontrivial one since the system has a "nonsummativity" character.

(3) *An estimation of the approximation* that such a decomposition may imply. Indeed, we are willing to make do with an approximate model if this is the price we must pay to control complexity. All we are then entitled to demand is that the degree of approximation remains known and, of course, tolerable.

Near-Complete Decomposability

A possibility of fulfilling these requirements is offered by the concept of *near-complete decomposability* and by its associated technique of aggregation of variables. The purpose of this monograph is to investigate in depth this possibility.

It is in economic theory that aggregation of variables has been most explicitly used as a technique to study and evaluate the dynamics of systems of great size and complexity. This technique is based on the idea that in many large systems all variables can somehow be clustered into a small number of groups so that: (i) the interactions among the variables of each single group may be studied as if interactions among groups did not exist, and (ii) interactions among groups may be studied without reference to the interactions within groups. This idea is of course quite general, and has, at least indirectly, been productive in disciplines other than economics; it is, for example, at the root of Boltzmann's fundamental hypothesis in statistical mechanics, of the Russell–Saunders approximation for the classification of atomic energy levels in quantum mechanics, and of Norton's and Thévenin's theorems in electric circuit synthesis.

This aggregation of variables yields rigorously correct results under two different types of conditions. The first type requires that interactions between groups of variables be independent from interactions within groups; then, these interactions between groups can be exactly analyzed without regard to the interactions within groups; the block stochastic systems studied in Chapter II and the lumpable Markov chain considered in Chapter III belong to the class of systems which satisfy the first type of conditions. The second type of conditions requires that variables be functions of variables of the same group only, i.e. that interactions between groups of variables be null. The system in this case can be said to be completely decomposable[1]: it truly consists of

[1] It is worth making the terminology more precise at the outset: such a system may be represented by a *completely decomposable* matrix, i.e., a square matrix such that an identical permutation of rows and columns leaves a set of square submatrices on the principal diagonal and zeros everywhere else; a *decomposable* matrix, as opposed to completely decomposable, is a matrix with zeros everywhere below the principal submatrices but not necessarily also above. *Near-complete decomposability* and *near-decomposability* are defined by replacing the zeros in these definitions by small nonzero numbers.

independent subsystems, each of which can be analyzed separately, without reference to the others.

Simon and Ando (1961) investigated circumstances under which variable aggregation still yields satisfactory approximations when interactions between groups of variables are nonnull and arbitrary, but weak compared with interactions within groups; such systems were qualified by Ando and Fisher (1963) as *nearly completely decomposable systems*. Several examples taken from economics (Simon and Ando, 1961), physics (Simon and Ando, 1961; Simon, 1962, 1969), and social sciences (Fisher and Ando, 1962) indicate that systems of that kind are likely to be more frequently encountered in reality than systems verifying the assumption of complete decomposability. This conjecture is based on Simon's proposition (Simon, 1962) that complex systems are often hierarchically organized and that hierarchies are likely to be nearly completely decomposable because interactions at the same level of a hierarchy are usually tighter than interactions between distinct levels.

Simon and Ando (1961) showed that aggregation of variables in nearly completely decomposable systems must separate the analysis of the short-run from that of the long-run dynamics. They proved two theorems. The first one states that, provided intergroup dependencies are sufficiently weak as compared to intragroup ones, in the short run the system will behave approximately, and can therefore be analyzed, as if it were completely decomposable. The second theorem states that even in the long run, when neglected intergroup dependencies have had time to influence the system behavior, the values of the variables within any group will remain approximately in the same ratio as if those intergroup influences had never existed. The results obtained in the short run will therefore remain approximately valid in the long run, as far as the *relative* behavior of the variables of the same group is concerned.

These two theorems will be formally introduced in Chapter I. We shall also indicate in that chapter how variables representing aggregated states can be used at an arbitrary number of levels of aggregation to evaluate the limiting behavior of stochastic systems that have a large number of states.

But the Simon and Ando theorems are in fact only existence theorems, which prove that whatever standard of approximation is required, a sufficient degree of near-complete decomposability always exists such that an analysis by aggregation can meet this standard. In Chapter II, we tackle the other end of the problem and seek means to determine from the characteristics of any given system which standard of approximation can be guaranteed when aggregation is used. This determination of the accuracy is of course an important issue if near-complete decomposability and variable aggregation are to become a modeling technique of some practical use: the results of an analysis remain useless if their degree of accuracy cannot be estimated. Our error analysis reveals that the approximation made by aggregation in nearly completely decomposable systems depends essentially on the degree of coupling between subsystems, and on the indecomposability of these subsystems. This study is based on elements of perturbation theory that have been developed by Wilkinson (1965).

Necessary and sufficient conditions for a system to enjoy the property of near-complete decomposability are examined in Chapter III. A sufficient condition expressed in terms of subsystem interdependency and indecomposability is established; this condition serves as a criterion for variable aggregation in the models analyzed in the following chapters.

The theory developed in the first part of this monograph applies only to stochastic systems; because of the unpredictability and randomness of the user demands upon computers, stochastic models are indeed among the more useful and realistic models for these systems. These theoretical developments can, however, be readily extended to other linear systems; moreover, the error analysis is new and its generality makes it useful for these other scientific areas mentioned above where near-complete decomposability has been observed and can be exploited.

Queueing Networks

The second part consists of three more chapters in which the concept of near-complete decomposability is exploited to analyze networks of interconnected queues. These networks are in several respects adequate models for studying congestion problems in computer systems (Wallace and Rosenberg, 1966; Smith, 1966; Arora and Gallo, 1971; Buzen, 1971a, b; Baskett *et al.*, 1975). They provide a formal context in which it is possible to bring out with sufficient rigor the basic aspects of near-complete decomposability in computer system structures. Moreover, we show that variable aggregation is in itself a useful technique to analyze these networks.

In Chapter IV we study the network model of Gordon and Newell (1967). An arbitrary fixed number N of customers make use of an arbitrary fixed number $L+1$ of resources, each of which provides a certain type of service. When service is completed by resource l, a customer proceeds directly to resource m with probability p_{lm}. The conditions under which such a system is nearly completely decomposable into L levels of aggregation are established in terms of the resource service rates and the transfer probabilities p_{lm}. It is shown in Chapter V that, when these conditions are fulfilled, the set of resources may be organized in a hierarchy of *aggregate resources*, each aggregate resource being analyzable merely as a single-server queueing system.

This approach presents several advantages, which are discussed in Chapter VI. First, the determination of the equilibrium marginal probabilities does not require, as in Gordon and Newell (1967), the inversion of the transfer matrix $[p_{lm}]$. Explicit and closed-form solutions are obtained for these probabilities, as well as for the average time a customer waits at each resource. These expressions facilitate the estimation of the equilibrium probabilities when the number of states is large, i.e., in the case of a large number of customers and/or resources. Next, these solutions apply to queueing networks that are more general than the model considered by Gordon and Newell (1967). They may be used to evaluate networks in which, as in the general model of Jackson (1963), the service rates are at each service stage almost arbitrary functions of the congestion at this stage. Transfer probabilities dependent on the congestions at the stages of departure and arrival may be taken into account as well. Moreover, this hierarchical model of aggregate resources appear to be a good approximation to multiqueue systems in which, instead of being exponentially distributed, the service times are random variables with arbitrary distribution functions. In other words, near-complete decomposability can under suitable circumstances dispense with the Poisson service time assumption, which is classically made in networks of queues by the sheer necessity of overcoming analytic difficulties.

In the remaining sections of Chapter VI we discuss other aspects of this analysis by resource aggregation: the possibility of using different modeling techniques at different levels of aggregation (heterogeneous aggregation), the evaluation of the network transient behavior, and the computational advantages of aggregation compared to other methods of network analysis. We show also in this chapter that load balancing and optimization of resource utilization are factors that promote near-complete decomposability, and thus hierarchical structuring, in queueing networks.

Computer System Models

Chapter VII opens the third part of this work, where the problem of computer system performance prediction, which originally motivated our study, is more concretely dealt with. The concepts and results introduced in the preceding chapters are applied to the analysis of different aspects, hardware and software, of the dynamic behavior of computer systems.

Hopefully, this last part should demonstrate that aggregation is not only an efficient technique for obtaining approximate results when numerous system parameters are involved, but also for gaining insight and conceptual clarity on the parts played by these parameters. The different models that are studied uncover several properties of the dynamics of the systems being analyzed, properties that would have been much more difficult to isolate without the distinction of short- and long-term dynamics in nearly completely decomposable systems.

We have already observed that hierarchical ordering and near-complete decomposability are two concepts that keep pace with each other.

Again, if near-complete decomposability is useful at all in computer systems, it is because the complexity of these systems, as for many artificial systems (Simon, 1962), frequently takes the form of hierarchy. Different reasons for this are discussed in these last chapters. Probably the most fundamental reason is that in computers the storage function itself is, by physical and economical necessity, hierarchically organized. A computer memory usually consists of a hierarchy of different storage media, of increasingly larger capacity, and correspondingly of increasingly slower access speed and lower cost per unit of storage. It would be impractical to equip a computer system with a single, huge storage medium: it would be either too slow or too expensive. This is true for present technologies and is likely to remain true for future ones.

This hierarchic organization of the storage function pervades many other functions of a computer system and makes near-complete decomposability an intrinsic property of many aspects of its dynamic behavior. For this reason, Chapter VII starts with the study of a nearly completely decomposable model of memory hierarchy that opportunely introduces several definitions and concepts needed later. In particular, considering a computer memory as a hierarchy of *aggregate memory levels*, performance criteria are defined that permit the evaluation of the benefit of multiprograming in memory hierarchies. Its ability to cope with state-dependent transfer probabilities renders this model also sensitive to allocation policies, which adjust dynamically to the demand of space allotted per program in each memory level.

The behavior of a machine is entirely determined by the program it executes, so that no performance evaluation can be done without realistic models of program behavior. In Chapter VIII a model of program behavior in paged-memory hierarchies is proposed. Existing models (Aho *et al.*, 1971; Spirn and Denning, 1972) are shown to give an incomplete account of the *locality property* of program behavior, that is, the property of programs to use for relatively long periods of time only small subsets of their data. We exploit the analogy between the concept of locality and the concept of aggregate to overcome this deficiency. This analogy leads to the combined use of a Markovian model of the program transitions between localities (transitions that reach a statistical equilibrium in the long term) and of separate models for the short-term behavior of the different localities. This combination gives, under the assumption of disjoint localities, better estimations of the efficiency of page replacement algorithms and of program behavior characteristics, such as the working set size distribution (Denning, 1968c). The conditions under which this last distribution is approximately normal and under which the assumptions of independent page references are valid are also clarified. The approach is illustrated by a numerical example, showing in particular that other models presented in the literature have computer time and space requirements that are beyond practical possibilities for programs of normal sizes.

Chapter IX is intended to illustrate the use of an aggregative model in analyzing a given complete computing system. We study the model of a time-sharing system in which memory space and processor time are respectively allocated on a demand paging and on a multiprograming basis. We first carry out the analysis of the short-term equilibrium attained by the page traffic between primary and secondary memory. Several interesting phenomena are then revealed by relating these results to the long-term behavior of the flow of transactions between the set of user consoles and the entire computing system. We show that, owing to the fluctuations of processor usage with the degree of multiprograming, the working conditions of a paging time-sharing system may be either *stable* or *unstable*. When the regime of operations is stable, the fluctuations of the congestion, measured by the current number of pending user transactions, are deadened as a result of the alterations they cause to the ratio of the transaction input rate to the processor usage factor. In the unstable regime, these congestion fluctuations are reinforced by the way they alter the input rate to processor usage ratio. We indicate how these stable and unstable values of the congestion may be calculated. Moreover, the analysis reveals that below a certain computing load the average congestion and the mean system response time increase less than linearly with the load, while beyond this computing load they increase more than linearly. This critical computing load is an extension to service systems with congestion-dependent service rate of the *saturation* point defined by Kleinrock (1968). These notions of instability and saturation lead to a more complete definition of the circumstances under which the severe system performance degradation known as *thrashing* (Denning, 1968b) may occur. They lead also to the definition of an optimum for the maximum degree of multiprograming.

The last chapter is a chapter of conjectures. These conjectures are drawn from the striking affinity between the concept of aggregate in nearly completely decomposable structures and the less precisely defined notion of building block, or module, so frequently invoked in computer system hardware and software engineering. More precisely, we discuss the similarity between the hierarchical model of aggregate resources defined in Chapter V and the structure of levels of abstraction advocated by Dijkstra (1969a) for the software of multiprograming computer systems. We conjecture that the criteria that indicate which resources should be aggregated are, in many cases, the same as those which specify what the levels of abstraction should be. The conditions for aggregation in such systems are expressed in terms of parameters closely related to the physical characteristics and the usage of the hardware resources. Hence, provided these parameters may be assessed, resource aggregation conditions might help the designer choose the ordering of the levels of abstraction. Variable aggregation appears in these two models as a technique that enables the level-by-level evaluation of the system to be interlaced with its level-by-level design. This stems from the fact that the conditions for aggregation are the same as those necessary to provide sufficiently approximate knowledge of the performances of a still incomplete system on which further design decisions may be based. Aggregative models of queue networks appear, in the sphere of computing system analytical models, as a counterpart of the level-by-level simulation techniques recommended by Parnas and Darringer (1967), Parnas (1969), Zurcher and Randell (1968), and Randell (1969).

Finally, in the conclusions, we recapitulate certain problems and questions that are raised and left open by this work.

CHAPTER I

Nearly Completely Decomposable Systems

This chapter formally introduces the basic assumptions supporting the property of near-complete decomposability and the two fundamental theorems that have been proved by Simon and Ando (1961). We limit our attention to the case of stochastic systems, which is the only case relevant to our further studies. Our developments deviate only slightly from those of Simon and Ando; their assumption of matrices with distinct roots is relaxed here and replaced by the assumption of matrices similar to matrices of diagonal form. We then proceed by indicating how aggregative variables can be used at an arbitrary number of levels of aggregation to evaluate the limiting state probability distribution of stochastic systems that enjoy the property of multilevel near-complete decomposability.

Almost all the basic mathematical notation used throughout the monograph is introduced in this chapter.

1.1 Eigencharacteristics and Condition Numbers

1.1.1 We shall assume all vectors to be column vectors and we shall use the symbol $\sim$ to denote the transpose operation. We are interested in stochastic systems of the form

$$\tilde{\mathbf{y}}(t+1) = \tilde{\mathbf{y}}(t)\mathbf{Q}, \tag{1.1}$$

where $\tilde{\mathbf{y}}(t)$ is a row probability vector and $\mathbf{Q}$ a stochastic matrix of order n. The system has n possible states, and $y_l(t)$ is the unconditional probability of the system being in the state l, $l = 1, \ldots, n$, at time t; the element q_{kl} of matrix $\mathbf{Q}$ is the conditional probability that the system is in the state l at time t given that it was in state k at time $t-1$.

Let us also consider a system

$$\tilde{\mathbf{y}}^*(t+1) = \tilde{\mathbf{y}}^*(t)\mathbf{Q}^*, \tag{1.2}$$

where the stochastic matrix $\mathbf{Q}^*$ is of order n and is equal to

$$\mathbf{Q}^* = \begin{bmatrix} \mathbf{Q}_1^* & & & & \\ & \ddots & & & \\ & & \mathbf{Q}_I^* & & \\ & & & \ddots & \\ & & & & \mathbf{Q}_N^* \end{bmatrix}.$$

The $\mathbf{Q}_I^*$ are square submatrices, and the remaining elements, not displayed, are all zero. Let $n(I)$ be the order of $\mathbf{Q}_I^*$; then $n = \sum_{I=1}^{N} n(I)$. The following notation is adopted to refer to the elements of the vector $\tilde{\mathbf{y}}^*(t)$:

$$\tilde{\mathbf{y}}^*(t) = [y_k{}^*(t)] = [[y_{i_1}^*(t)] \cdots [y_{i_I}^*(t)] \cdots [y_{i_N}^*(t)]],$$

where $[y_{i_I}^*(t)]$ is a vector of elements of $[y_k{}^*(t)]$, so that if

$$y_{i_I}^*(t) = y_k{}^*(t),$$

then

$$k = \sum_{J=1}^{I-1} n(J) + i.$$

The matrix $\mathbf{Q}^*$ is said to be *completely decomposable*. It is clear that in the system

$$\tilde{\mathbf{y}}^*(t) = \tilde{\mathbf{y}}^*(0)\,\mathbf{Q}^{*t}$$

the subvector $[y_{i_I}^*(t)]$ depends, for any t, only on $[y_{i_I}^*(0)]$ and $\mathbf{Q}_I^*$, and is independent of $[y_{i_J}^*(0)]$ and $\mathbf{Q}_J^*$, $J \neq I$.

We shall now assume that both the rows and columns of $\mathbf{Q}$ can be arranged by the same appropriate permutation, so that

$$\mathbf{Q} = \mathbf{Q}^* + \varepsilon\mathbf{C}, \tag{1.3}$$

where $\mathbf{C}$ is a square matrix of the same order as $\mathbf{Q}^*$, which has the property of keeping both $\mathbf{Q}$ and $\mathbf{Q}^*$ stochastic, and where ε is a real positive number, small compared to the elements of $\mathbf{Q}^*$. Matrices of the form of $\mathbf{Q}$ are defined by Ando and Fisher (1963) as being *nearly completely decomposable matrices.* Roughly speaking, they consist of principal submatrices, most elements of which are substantially larger than the elements outside these submatrices.

The row and column partition of $\mathbf{Q}^*$ will be applied to $\mathbf{Q}$, $\mathbf{C}$, and all later matrices: we denote as $\mathbf{Q}_{IJ}$ the submatrix of $\mathbf{Q}$ at the intersection of the Ith set of rows and the Jth set of columns, and as $q_{i_I j_J}$ the element at the intersection of the ith row and jth column of $\mathbf{Q}_{IJ}$.

Since the matrices $\mathbf{Q}$ and $\mathbf{Q}_I^*$, $I = 1, \ldots, N$, are stochastic, the row sums of $\mathbf{C}$ are equal to zero. We choose nonpositive values for all elements $c_{i_I k_I}$ and nonnegative values for all elements $c_{i_I k_J}$, $I \neq J$, so that we have for each row i_I

$$\sum_{k=1}^{n(I)} c_{i_I k_I} = -\sum_{J=1, J\neq I}^{N} \sum_{j=1}^{n(J)} c_{i_I j_J}.$$

Moreover, we choose ε and $\mathbf{C}$ so that for all rows i_I

$$\varepsilon \sum_{J=1, J\neq I}^{N} \sum_{j=1}^{n(J)} c_{i_I j_J} = \sum_{J=1, J\neq I}^{N} \sum_{j=1}^{n(J)} q_{i_I j_J},$$

and

$$\varepsilon = \max_{i_I} \left(\sum_{J=1, J\neq I}^{N} \sum_{j=1}^{n(J)} q_{i_I j_J} \right). \tag{1.4}$$

Thus,

$$\max_{i_I} \left(\sum_{J=1, J\neq I}^{N} \sum_{j=1}^{n(J)} c_{i_I j_J} \right) = -\max_{i_I} \left(\sum_{k=1}^{n(I)} c_{i_I k_I} \right) = 1,$$

and $|c_{i_I j_J}| \leqslant 1$ for all i_I, j_J.

We shall sometimes refer to ε as *the maximum degree of coupling between subsystems* $\mathbf{Q}_{II}$.

An example of matrices $\mathbf{Q}$, $\mathbf{Q}^*$, and $\mathbf{C}$ is

$$\mathbf{Q} = \begin{bmatrix} 0.5 & 0.45 & 0.05 \\ 0.6 & 0.375 & 0.025 \\ 0.025 & 0.025 & 0.95 \end{bmatrix},$$

with $\mathbf{Q} = \mathbf{Q}^* + \varepsilon\mathbf{C}$ being given by

$$\mathbf{Q} = \begin{bmatrix} 0.5 & 0.5 & 0 \\ 0.625 & 0.375 & 0 \\ 0 & 0 & 1 \end{bmatrix} + 5 \times 10^{-2} \begin{bmatrix} 0 & -1 & 1 \\ -0.5 & 0 & 0.5 \\ 0.5 & 0.5 & -1 \end{bmatrix}$$

with

$$\mathbf{Q}_1^* = \begin{bmatrix} 0.5 & 0.5 \\ 0.625 & 0.375 \end{bmatrix} \quad \text{and} \quad \mathbf{Q}_2^* = [1].$$

1.1.2 Each stochastic matrix $\mathbf{Q}_I^*$, $I = 1, \ldots, N$, is supposed to be *indecomposable*, i.e., there exists no permutation matrix $\mathbf{T}$ such that

$$\mathbf{T}\mathbf{Q}_I^*\tilde{\mathbf{T}} = \begin{bmatrix} \mathbf{A} & \mathbf{B} \\ \mathbf{0} & \mathbf{D} \end{bmatrix},$$

where $\mathbf{A}$ and $\mathbf{D}$ are square and $\mathbf{0}$ is a zero matrix. Since $\mathbf{Q}_I^*$ is a square nonnegative indecomposable matrix, it results from the Perron–Frobenius theorem (see, e.g., Marcus and Minc, 1964, p. 124) that its largest eigenvalue in module is simple; as $\mathbf{Q}_I^*$ is stochastic, this maximal eigenvalue is equal to unity.

We use $\lambda^*(i_I)$, $i = 1, \ldots, n(I)$, to denote the eigenvalues of $\mathbf{Q}_I^*$, and we suppose that they are ordered so that

$$\lambda^*(1_I) = 1 > |\lambda^*(2_I)| \geqslant |\lambda^*(3_I)| \geqslant \cdots \geqslant |\lambda^*(n(I)_I)|.$$

The nonmaximal eigenvalues need not be distinct.

Let us now define δ^* as being the minimum of the absolute values of the differences between unity and all eigenvalues of $\mathbf{Q}^*$ that are not unity. We have

$$|1-\lambda^*(i_I)| \geqslant \delta^* > 0, \qquad i = 2, \ldots, n(I), \quad I = 1, \ldots, N. \tag{1.5}$$

The eigenvalues of a matrix being continuous functions of its elements (see, for example, Coolidge, 1959), we can define for any positive real number δ, however small, a small enough ε so that, for every eigenvalue $\lambda^*(i_I)$ of $\mathbf{Q}^*$, there exists an eigenvalue $\lambda(i_I)$ of $\mathbf{Q}$ such that for all i_I

$$|\lambda(i_I) - \lambda^*(i_I)| < \delta. \tag{1.6}$$

Hence, we may classify the eigenvalues of $\mathbf{Q}$ so that

$$\begin{aligned} &|1-\lambda(1_I)| < \delta, && I = 1, \ldots, N, \\ &|1-\lambda(i_I)| > \delta^* - \delta, && I = 1, \ldots, N, \quad i = 2, \ldots, n(I), \end{aligned} \tag{1.7}$$

where δ approaches zero with ε.

Again we assume that $\mathbf{Q}$ is indecomposable, and thus that $\lambda(1_1) = 1$ is a simple eigenvalue; the other eigenvalues of $\mathbf{Q}$ need not be simple.

1.1.3 For the sake of simplicity, the subsequent analysis will be restricted to the case where all elementary divisors of $\mathbf{Q}$ and $\mathbf{Q}^*$ are linear. Then we can assume the existence of *complete* sets of left and right eigenvectors for these matrices.

We use $\tilde{\mathbf{v}}^*(i_I)$ and $\mathbf{\upsilon}^*(i_I)$, respectively, to denote the left and right eigenvectors of $\mathbf{Q}^*$ corresponding to $\lambda^*(i_I)$. These vectors, of length n, are obtained by bordering with appropriate numbers of zeros both extremities of the corresponding eigenvectors of $\mathbf{Q}_I^*$. These vectors are supposed to be normalized so that

$$\forall i_I: \quad \|\mathbf{v}^*(i_I)\|_1 = \|\mathbf{\upsilon}^*(i_I)\|_1 = 1, \tag{1.8a}$$

where the vector norm $\|\mathbf{y}\|_1$ is defined as

$$\|\mathbf{y}\|_1 = \sum_i |y_i|.$$

The vectors that correspond to a multiple eigenvalue $\lambda^*(i_I)$ are not uniquely determined; we shall assume in this case that, if $\lambda^*(i_I)$ is of multiplicity m, a choice of m independent vectors $\mathbf{v}^*(i_I)$ and $\mathbf{\upsilon}^*(i_I)$ is made.

Since $\mathbf{Q}_I{}^*$ is stochastic, all elements of the right eigenvector $\mathfrak{v}^*(1_I)$ are equal. Thus, to satisfy (1.8a) we must have for all i_I

$$v_{i_I}^*(1_I) = n(I)^{-1}. \tag{1.8b}$$

Similarly for $\mathbf{Q}$, the left and right eigenvectors associated with $\lambda(i_I)$ will be denoted $\tilde{\mathbf{v}}(i_I)$ and $\mathfrak{v}(i_I)$, respectively, a particular choice being made if $\lambda(i_I)$ is not simple; they are supposed to have same normalization as in (1.8a), so that we have for all i_I

$$v_{i_I}(1_1) = n^{-1}.$$

Finally, we shall make repeated use, for the matrices $\mathbf{Q}$ and $\mathbf{Q}^*$, of the scalars $(i = 1, \ldots, n(I),\ I = 1, \ldots, N)$

$$s(i_I) = \tilde{\mathbf{v}}(i_I)\,\mathfrak{v}(i_I), \qquad s^*(i_I) = \tilde{\mathbf{v}}^*(i_I)\,\mathfrak{v}^*(i_I), \tag{1.9a}$$

where it is understood that the vectors are normalized. These scalars are the *condition numbers* of $\mathbf{Q}$ and $\mathbf{Q}^*$, respectively (Wilkinson, 1965, p. 89).

Since the matrices $\mathbf{Q}$ and $\mathbf{Q}_I{}^*$, $I = 1, \ldots, N$, are stochastic, we have

$$s(1_1) = n^{-1}, \qquad s^*(1_I) = n(I)^{-1}. \tag{1.9b}$$

More generally, we have also

$$|s(i_I)| = |\tilde{\mathbf{v}}(i_I)\,\mathfrak{v}(i_I)| \leqslant \|\mathbf{v}(i_I)\|_1 \|\mathfrak{v}(i_I)\|_1 = 1. \tag{1.10}$$

Relations (1.10) apply to $s^*(i_I)$.

1.2 The Simon–Ando Theorems

We assumed that $\mathbf{Q}$ has linear elementary divisors, and thus that complete sets of left and right eigenvectors (not necessarily unique if the eigenvalues are not simple) exist such that

$$\tilde{\mathbf{v}}(i_I)\,\mathfrak{v}(j_J) = 0, \qquad i_I \neq j_J, \tag{1.11}$$

and, from the definition of the condition numbers,

$$\tilde{\mathbf{v}}(i_I)\,s(i_I)^{-1}\mathfrak{v}(i_I) = 1. \tag{1.12}$$

The Jordan canonical form of $\mathbf{Q}$ is thus diagonal. A nonsingular matrix $\mathbf{H}$ exists such that

$$\mathbf{Q} = \mathbf{H}\,\mathbf{diag}(\lambda(i_I))\,\mathbf{H}^{-1}, \tag{1.13}$$

where the i_Ith column of $\mathbf{H}$ is the vector $s(i_I)^{-1}\mathfrak{v}(i_I)$, the i_Ith row of $\mathbf{H}^{-1}$ is the left eigenvector $\tilde{\mathbf{v}}(i_I)$, $i = 1, \ldots, n(I)$, $I = 1, \ldots, N$, and where $\mathbf{diag}(\lambda(i_I))$

is a diagonal matrix whose elements are the roots of $\mathbf{Q}$:

$$\mathbf{diag}(\lambda(i_I)) = \begin{bmatrix} \lambda(1_1) & & & & \\ & \lambda(2_1) & & & \\ & & \ddots & & \\ & & & \lambda(i_I) & \\ & & & & \ddots \\ & & & & & \lambda(n(N)_N) \end{bmatrix}.$$

Clearly, from (1.13)

$$\mathbf{Q}^t = \mathbf{H}\,\mathbf{diag}(\lambda^t(i_I))\,\mathbf{H}^{-1}.$$

Thus, from the definition of $\mathbf{H}$ and $\mathbf{H}^{-1}$,

$$\mathbf{Q}^t = \sum_{I=1}^{N} \sum_{i=1}^{n(I)} \lambda^t(i_I)\, s(i_I)^{-1} \mathbf{v}(i_I)\, \tilde{\mathbf{v}}(i_I).$$

Introducing the matrices

$$\mathbf{Z}(i_I) = s(i_I)^{-1} \mathbf{v}(i_I)\, \tilde{\mathbf{v}}(i_I), \qquad i = 1, \ldots, n(I), \quad I = 1, \ldots, N,$$

we obtain

$$\mathbf{Q}^t = \sum_{I=1}^{N} \sum_{i=1}^{n(I)} \lambda^t(i_I)\, \mathbf{Z}(i_I), \tag{1.14}$$

which is the spectral decomposition of $\mathbf{Q}$.

It results from their definition that the matrices $\mathbf{Z}(i_I)$ have the following properties:

$$\mathbf{Z}(i_I)\,\mathbf{Z}(i_I) = \mathbf{Z}(i_I) \qquad \text{(idempotency)},$$

$$\mathbf{Z}(i_I)\,\mathbf{Z}(j_J) = \mathbf{0}_n, \qquad i_I \neq j_J \quad \text{(orthogonality)},$$

$$\sum_{I=1}^{N} \sum_{i=1}^{n(I)} \mathbf{Z}(i_I) = \mathbf{I}_n,$$

$$\operatorname{tr} \mathbf{Z}(i_I) = 1,$$

where $\mathbf{I}_n$ denotes the $n \times n$ unit matrix, $\mathbf{0}_n$ the $n \times n$ zero matrix, and $\operatorname{tr} \mathbf{Z}$ the trace of $\mathbf{Z}$.

Following the classification of eigenvalues defined by (1.7), we now divide the right-hand side of (1.14) into terms

$$\mathbf{Q}^t = \mathbf{Z}(1_1) + \sum_{I=2}^{N} \lambda^t(1_I)\, \mathbf{Z}(1_I) + \sum_{I=1}^{N} \sum_{i=2}^{n(I)} \lambda^t(i_I)\, \mathbf{Z}(i_I). \tag{1.15}$$

Similarly for $\mathbf{Q}^*$ we have the matrices

$$\mathbf{Z}^*(i_I) = s^*(i_I)^{-1} \mathbf{v}^*(i_I)\, \tilde{\mathbf{v}}^*(i_I),$$

and we can isolate in the spectral decomposition of $\mathbf{Q}^{*t}$ the roots $\lambda^*(1_I)$, $I = 1, \ldots, N$, which are equal to unity:

$$\mathbf{Q}^{*t} = \sum_{I=1}^{N} \mathbf{Z}^*(1_I) + \sum_{I=1}^{N} \sum_{i=2}^{n(I)} \lambda^{*t}(i_I)\,\mathbf{Z}^*(i_I). \tag{1.16}$$

The time behaviors of $\tilde{\mathbf{y}}(t)$ and $\tilde{\mathbf{y}}^*(t)$ that have been defined by (1.1) and (1.2), respectively, are also specified by (1.15) and (1.16). The comparison of these behaviors, which is made in the next section, is based on the following theorems proven by Simon and Ando (1961):

Theorem 1.1 *For an arbitrary positive real number ξ, there exists a number ε_ξ such that for $\varepsilon < \varepsilon_\xi$,*

$$\max_{k,l} |z_{kl}(i_I) - z^*_{kl}(i_I)| < \xi$$

with

$$2 \leqslant i \leqslant n(I), \qquad 1 \leqslant I \leqslant N, \qquad 1 \leqslant k, l \leqslant n.$$

Theorem 1.2 *For an arbitrary positive real number ω, there exists a number ε_ω such that for $\varepsilon < \varepsilon_\omega$,*

$$\max_{i,j} |z_{i_I j_J}(1_K) - v^*_{j_J}(1_J)\,\alpha_{IJ}(1_K)| < \omega$$

with

$$1 \leqslant K, I, J \leqslant N, \qquad 1 \leqslant i \leqslant n(I), \qquad 1 \leqslant j \leqslant n(J),$$

and where $\alpha_{IJ}(1_K)$ is given by

$$\alpha_{IJ}(1_K) = \sum_{i=1}^{n(I)} \sum_{j=1}^{n(J)} v^*_{i_I}(1_I)\, z_{i_I j_J}(1_K). \tag{1.17}$$

We omit the proofs of these theorems, as the method of proof has little relevance to the developments we make in later chapters.

1.3 Interpretation of Theorems

The implications of these two theorems may be discussed on rather intuitive grounds.

Since, by (1.7), the $\lambda(1_I)$, $I = 1, \ldots, N$, are close to unity, for any small t (say t smaller than some time T_2) $\lambda^t(1_I)$, $I = 1, \ldots, N$, will also stay close to unity. Therefore, the first two terms of the right-hand side of (1.15) will not vary very much for $t < T_2$ while the first term of the right-hand side of (1.16) will not vary at all. Thus, for $t < T_2$, the time behaviors of $\mathbf{y}(t)$ and $\mathbf{y}^*(t)$ are defined by the last terms of $\mathbf{Q}^t$ and $\mathbf{Q}^{*t}$, respectively. But as $\varepsilon \to 0$, it results from (1.6) that

$$\lambda(i_I) \to \lambda^*(i_I), \tag{1.18}$$

and from Theorem 1.1 that

$$\mathbf{Z}(i_I) \to \mathbf{Z}^*(i_I), \qquad i = 2, \ldots, n(I), \quad I = 1, \ldots, N.$$

Hence, for ε sufficiently small and $t < T_2$, the time path of $\mathbf{y}(t)$ must be very close to that of $\mathbf{y}^*(t)$.

Now, since the $\lambda^*(i_I)$, $i = 2, \ldots, n(I)$, $I = 1, \ldots, N$, are less than unity [from (1.5)], for any positive real ξ_1 we can define a smallest time interval T_1^* such that

$$\max_{\substack{k,l \\ 1 \leqslant k,l \leqslant n}} \left| \sum_{I=1}^{N} \sum_{i=2}^{n(I)} \lambda^{*t}(i_I)\, z_{kl}^*(i_I) \right| < \xi_1, \qquad \text{for} \quad t > T_1^*.$$

Likewise, we can define a smallest time interval T_1 such that

$$\max_{\substack{k,l \\ 1 \leqslant k,l \leqslant n}} \left| \sum_{I=1}^{N} \sum_{i=2}^{n(I)} \lambda^{t}(i_I)\, z_{kl}(i_I) \right| < \xi_1, \qquad \text{for} \quad t > T_1.$$

Moreover, (1.18) and Theorem 1.1 ensure that

$$T_1 \to T_1^* \qquad \text{as} \quad \varepsilon \to 0,$$

T_1^* being independent of ε. Since T_2 can be made as large as we want by taking ε sufficiently small, while T_1^* remains independent of ε, *we can choose a small enough ε so that T_2 is larger than T_1.*

Finally, provided that ε is not identically zero so that λ_{1_I}, $I = 2, \ldots, N$, is not identically unity, there will come a time $T_3 > T_2$ such that for a small enough positive real number ξ_3,

$$\max_{\substack{k,l \\ 1 \leqslant k,l \leqslant n}} \left| \sum_{I=2}^{N} \lambda^{t}(1_I)\, z_{kl}(1_I) \right| < \xi_3.$$

Time T_3, like T_2, increases without limit as $\varepsilon \to 0$. For $T_2 < t < T_3$, the last term of $\mathbf{Q}^t$ is negligible and the time path of $\mathbf{y}(t)$ is determined by the first two terms of $\mathbf{Q}^t$. But Theorem 1.2 specifies that for any I and J the elements of $\mathbf{Z}(1_K)$,

$$z_{i_I 1_J}(1_K), \ldots, z_{i_I j_J}(1_K), \ldots, z_{i_I n(J)_J}(1_K)$$

depend essentially upon I, J, and j, and are almost independent of i. That is, for any I and J they are proportional to the elements of the steady-state vector of $\mathbf{Q}_J^*$

$$v_{1_J}^*(1_J), \ldots, v_{j_J}^*(1_J), \ldots, v_{n(J)_J}^*(1_J) \tag{1.19}$$

and are approximately the same for $i = 1, \ldots, n(I)$.

Thereby, since for $T_2 < t < T_3$, $\mathbf{Q}^t$ is mainly determined by the first two terms of (1.15), the vector $\mathbf{y}(t)$ will vary with t during that period keeping among its elements $y_{j_J}(t)$ of every subset J an approximately constant *ratio* that is identical to the ratio between the elements of (1.19).

Finally, for $t > T_3$, all terms of $\mathbf{Q}^t$ except the first one become negligible and the behavior of $\mathbf{y}(t)$ is dominated by the largest root of $\mathbf{Q}$, while $\mathbf{Q}$ as a whole evolves toward its steady state defined by the unique vector $\mathbf{v}(1_1)$.

We may summarize the above discussion by saying that the dynamic behavior of a system representable by a nearly completely decomposable matrix may be dissociated into four stages called by Simon and Ando (1961), respectively, (i) short-term dynamics, (ii) short-term equilibrium, (iii) long-term dynamics, (iv) long-term equilibrium. More precisely, these stages are:

(i) Short-term dynamics: $t < T_1$. The preponderantly varying term of $\mathbf{Q}^t$ is the last one, but this term is close to the last one of $\mathbf{Q}^{*t}$, $\mathbf{y}(t)$ and $\mathbf{y}^*(t)$ evolve similarly.

(ii) Short-term equilibrium: $T_1 < t < T_2$. The last terms of $\mathbf{Q}^t$ and $\mathbf{Q}^{*t}$ have vanished while the time powers of the N predominant eigenvalues $\lambda^t(1_I)$, $I = 1, \ldots, N$, remain close to unity. A similar equilibrium is being reached within each subsystem of $\mathbf{Q}$ and $\mathbf{Q}^*$.

(iii) Long-term dynamics: $T_2 < t < T_3$. The preponderantly varying term of $\mathbf{Q}^t$ is the second one. The whole nearly completely decomposable system moves toward equilibrium, but the short-term equilibrium relative values of the variables within each subsystem are approximately maintained.

(iv) Long-term equilibrium: $T_3 < t$. The first term $\mathbf{Q}^t$ dominates all the others. A global equilibrium is attained in the whole system.

To conclude, it is possible to specify for a nearly completely decomposable system a short-term period during which it behaves approximately as if it were completely decomposable, each subsystem approaching a local equilibrium approximately independently of the others. The length of this short-term period and the degree of approximation are of course related; this relation will be more closely examined in Chapter II. Finally, and this is probably the more important result obtained by Simon and Ando (1961), after this short-term period is over, these local equilibrium states remain approximately maintained in each subsystem during the period when the system, as a whole, evolves toward a global equilibrium under the influence of the weak interactions among subsystems.

1.4 Aggregation of Variables

1.4.1 An important consequence of this distinction between short- and long-term dynamics is the justification of the use of a process of variable aggregation to analyze the time behavior of a nearly completely decomposable system. Alternatively, near-complete decomposability may be considered as giving criteria according to which variables should be aggregated in a system. Let us clarify these points.

The problem of aggregation may be summarized as follows. The feasibility of aggregation in a linear $n \times n$ system of the form

$$\tilde{\mathbf{y}}(t+1) = \tilde{\mathbf{y}}(t)\mathbf{Q} \tag{1.20}$$

depends on the existence of

(i) N "aggregating functions"

$$Y_I(t) = \mathscr{A}_I(y_i(t)), \qquad I = 1, \ldots, N,$$

(ii) n "disaggregating functions"

$$y_i(t) = \mathscr{D}_i(Y_I(t)), \qquad i = 1, \ldots, n, \tag{1.21}$$

(iii) a new $N \times N$ system of relations

$$\tilde{\mathbf{Y}}(t+1) = \tilde{\mathbf{Y}}(t)\mathbf{P}, \tag{1.22}$$

the elements of $\mathbf{P}$ being functions of those of $\mathbf{Q}$.

N is hopefully much smaller than n, while the elements of $\mathbf{P}$ and the disaggregating functions are, ideally, rather easy functions to compute. Moreover, the $N \times N$ system $\mathbf{P}$ and the disaggregating functions must be such that the time behavior of the $y_i(t)$, as yielded by (1.21) and (1.22), must be an acceptable approximation to the time behavior defined by (1.20). Aggregation is basically a means of simplifying the analysis by breaking it up into stages at which the systems like (1.21) and (1.22) that must be resolved are of lesser complexity and smaller size.

1.4.2 ***Eigenvector Approximation*** Let us now assume that the system (1.20) is stochastic and nearly completely decomposable into a set of subsystems $\mathbf{Q}_{II}$, $I = 1, \ldots, N$. Then it results from the second Simon–Ando theorem that a time T_2 exists after which the probabilities $y_{i_I}(t)$ of each group I move in proportion to each other, keeping among themselves approximately the same ratio as the elements of the steady-state vector $\mathbf{v}^*(1_I)$ of $\mathbf{Q}_I{}^*$.

Thus, if we define

$$Y_I(t) = \sum_{i=1}^{n(I)} y_{i_I}(t), \tag{1.23}$$

for $t > T_2$ we may write

$$y_{i_I}(t)/Y_I(t) \simeq v_{i_I}^*(1_I). \tag{1.24}$$

On the other hand, the probability that the system defined by (1.20) will be at any time $t+1$ in any one state $j_J, j = 1, \ldots, n(J)$, of group J, on condition that it was in any one state i_I, $i = 1, \ldots, n(I)$, of group I at time t, is given by

$$p_{IJ}(t+1) = (Y_I(t))^{-1} \sum_{i=1}^{n(I)} y_{i_I}(t) \sum_{j=1}^{n(J)} q_{i_I j_J}.$$

Thus, for $t > T_2$ we have approximately

$$p_{IJ}(t+1) \simeq p_{IJ} = \sum_{i=1}^{n(I)} v_{i_I}^*(1_I) \sum_{j=1}^{n(J)} q_{i_I j_J}. \qquad (1.25)$$

The time-independent matrix $\mathbf{P} = [p_{IJ}]$ is stochastic and the elements, say $X_I(1)$, of its steady-state probability vector $\mathbf{X}(1)$,

$$\mathbf{X}(1)(\mathbf{P}-\mathbf{I}_N) = \mathbf{0} \qquad (1.26)$$

are good approximations (for $t > T_2$) to the probabilities $Y_I(t)$ of being in any one state of subset I. Moreover, since for $t \to \infty$, $y_{i_I}(t) \to v_{i_I}(1_1)$, (1.23) and (1.24) indicate that variables $x_{i_I}(1)$ defined by the disaggregating functions

$$x_{i_I}(1) = X_I(1)\, v_{i_I}^*(1_I), \qquad i = 1, \ldots, n(I), \quad I = 1, \ldots, N, \qquad (1.27)$$

are good approximations to the steady-state probabilities $v_{i_I}(1_1)$.

The obvious advantage of this approach is that it reduces the resolution of the linear system of order n

$$\tilde{\mathbf{v}}(1_1)(\mathbf{Q}-\mathbf{I}_n) = \mathbf{0}$$

to the resolution of the N small $n(I) \times n(I)$ independent systems

$$\tilde{\mathbf{v}}^*(1_I)(\mathbf{Q}_I{}^* - \mathbf{I}_{n(I)}) = \mathbf{0}, \qquad I = 1, \ldots, N,$$

and of the $N \times N$ system (1.26).

Yet the $x_{i_I}(1)$ are but approximations to the $v_{i_I}(1_1)$. The purpose of Chapter II is to determine how good these approximations are. We shall also indicate there how the other eigenvectors of $\mathbf{Q}$ can be approximated by a similar process of aggregation.

1.4.3 ***Approximation of the Inverse*** It is possible to relate the transition probabilities p_{IJ} to the coefficients α_{IJ} introduced in the second Simon–Ando theorem. We have

$$p_{IJ} = \sum_{i=1}^{n(I)} v_{i_I}^*(1_I) \sum_{j=1}^{n(J)} q_{i_I j_J}.$$

With the idempotent expansion of $\mathbf{Q}$, we obtain

$$p_{IJ} = \sum_{i=1}^{n(I)} v_{i_I}^*(1_I) \left[\sum_{K=1}^{N} \lambda(1_K) \sum_{j=1}^{n(J)} z_{i_I j_J}(1_K) + \sum_{K=1}^{N} \sum_{k=2}^{n(K)} \lambda(k_K) \sum_{j=1}^{n(J)} z_{i_I j_J}(k_K) \right].$$

From Theorem 1.1 we know that $z_{i_I j_J}(k_K) \to z^*_{i_I j_J}(k_K)$ for $k \neq 1$ and ε sufficiently small, and that these matrices $\mathbf{Z}^*(k_K)$, $k \neq 1$, are orthogonal to $\mathbf{Z}^*(1_K)$, which has precisely the vectors $\mathbf{v}^*(1_I)$ as its rows where the elements are not zero; thus for all I, J, and $\varepsilon \to 0$,

$$p_{IJ} \simeq \sum_{i=1}^{n(I)} v^*_{i_I}(1_I) \sum_{K=1}^{N} \lambda(1_K) \sum_{j=1}^{n(J)} z_{i_I j_J}(1_K),$$

which, because of (1.17), yields

$$P_{IJ} \simeq \sum_{K=1}^{N} \lambda(1_K)\,\alpha_{IJ}(1_K). \tag{1.28}$$

This relation calls for the following comment. A third theorem by Simon and Ando (1961) on stochastic nearly completely decomposable matrices states improperly that the right-hand side of (1.28) is the idempotent expansion of an $N \times N$ matrix. This would imply that the set of matrices $\mathbf{A}(1_K) = [\alpha_{IJ}(1_K)]$, $K = 1, \ldots, N$, must have exactly the same properties of orthogonality and idempotency as the matrices $\mathbf{Z}(1_K)$ from which they are derived. This is not the case. For $I, J = 1, \ldots, N$, we have only

$$\begin{aligned} &\lim_{\varepsilon \to 0} [\mathbf{A}(1_K) \times \mathbf{A}(1_K) - \mathbf{A}(1_K)] = \mathbf{0}_N && \text{(idempotency)}, \\ &\lim_{\varepsilon \to 0} \mathbf{A}(1_K) \times \mathbf{A}(1_L) = \mathbf{0}_N, \qquad K \neq L && \text{(orthogonality)}, \\ &\lim_{\varepsilon \to 0} \sum_{K=1}^{N} \mathbf{A}(1_K) = \mathbf{I}_N. \end{aligned} \tag{1.29}$$

The first two limits result from the corresponding properties of the matrices $\mathbf{Z}(k_K)$ and from the approximation defined by Theorem 1.2. Indeed we have

$$\begin{aligned} \sum_{J=1}^{N} \alpha_{IJ}(1_K)\alpha_{JH}(1_L) &= \sum_{J=1}^{N} \alpha_{IJ}(1_K) \sum_{j=1}^{n(J)} \sum_{h=1}^{n(H)} v^*_{j_J}(1_J)\, z_{j_J h_H}(1_L) \\ &= \sum_{h=1}^{n(H)} \sum_{J=1}^{N} \sum_{j=1}^{n(J)} \alpha_{IJ}(1_K)\, v^*_{j_J}(1_J)\, z_{j_J h_H}(1_L) \\ &\simeq \sum_{h=1}^{n(H)} \sum_{J=1}^{N} \sum_{j=1}^{n(J)} z_{i_I j_J}(1_K)\, z_{j_J h_H}(1_L). \end{aligned}$$

For $K \neq L$, this sum is clearly zero since the matrices $\mathbf{Z}(1_K)$ and $\mathbf{Z}(1_L)$ are orthogonal: hence the second limit in (1.29); and since $\mathbf{Z}(1_K)$ is idempotent, for $L = K$ this sum becomes approximately

$$\sum_{h=1}^{n(H)} z_{i_I h_H}(1_K).$$

By Theorem 1.2 this sum is approximately

$$\sum_{h=1}^{n(H)} v^*_{h_H}(1_H)\,\alpha_{IH}(1_K) = \alpha_{IH}(1_K),$$

which proves the first limit in (1.29).

The third limit in (1.29) is an immediate result of (1.28) if one notes that $\lambda(1_K) \to 1$ and $\mathbf{P} \to \mathbf{I}_N$ when $\varepsilon \to 0$.

These limits (1.29) show that the set of matrices $\mathbf{A}(1_K)$ converges toward an orthogonal basis *only asymptotically as* $\varepsilon \to 0$.

On the other hand, from relation (1.28) we obtain

$$\lim_{\varepsilon \to 0} \left[\mathbf{P} - \sum_{K=1}^{N} \lambda(1_K) \mathbf{A}(1_K) \right] = \mathbf{0}_N. \tag{1.30}$$

Because of this limiting behavior (1.29) and (1.30) the approximation of the inverse $\mathbf{Q}^{-1}$ proposed by Simon and Ando (1961) is possible. Indeed from (1.15) we have

$$\mathbf{Q}^{-1} = \sum_{K=1}^{N} \lambda(1_K)^{-1} \mathbf{Z}(1_K) + \sum_{K=1}^{N} \sum_{k=2}^{n(K)} \lambda(k_K)^{-1} \mathbf{Z}(k_K). \tag{1.31}$$

From Theorem (1.1) we obtain

$$\mathbf{Q}^{-1} \simeq \sum_{K=1}^{N} \lambda(1_K)^{-1} \mathbf{Z}(1_K) + \mathbf{Q}^{*-1} - \sum_{K=1}^{N} \mathbf{Z}^*(1_K). \tag{1.32}$$

However, by Theorem (1.2) we can write

$$\sum_{K=1}^{N} \lambda(1_K)^{-1} z_{i_I j_J}(1_K) \simeq v_{j_J}^*(1_J) \sum_{K=1}^{N} \lambda(1_K)^{-1} \alpha_{IJ}(1_K),$$

and by (1.29) and (1.30)

$$\sum_{K=1}^{N} \lambda(1_K)^{-1} \alpha_{IJ}(1_K) \simeq (P^{-1})_{IJ},$$

where $(P^{-1})_{IJ}$ is the (I, J)th element of $\mathbf{P}^{-1}$. Thus we have

$$\sum_{K=1}^{N} \lambda(1_K)^{-1} z_{i_I j_J}(1_K) \simeq v_{j_J}^*(1_J)(P^{-1})_{IJ}. \tag{1.33}$$

If we remember that each matrix $\mathbf{Z}^*(1_K)$ has all its rows identical to the first eigenvector $\mathbf{v}^*(1_K)$ of $\mathbf{Q}_K^*$, we see that (1.32) gives a useful approximation to the inverse of a matrix $\mathbf{Q}$. It reduces the computation of the inverse of an $n \times n$ matrix, $n = \sum_{I=1}^{N} n(I)$, of type $\mathbf{Q}$ to the computation of the inverse of an $N \times N$ aggregative matrix $\mathbf{P}$, and to the computation of the inverse and of the first eigenvector of the N smaller $n(I) \times n(I)$ matrices $\mathbf{Q}_I^*$. This is another example of the potentialities of an aggregation method.

1.4.4 ***Remarks*** A concept related to the concept of aggregation has been used by van Emden (1969) to analyze Markov chains. States of the chain are "fused," the result being a chain again with fewer states. Entropy is used as a criterion to determine which states should be clustered. Kemeny and Snell

(1960) speak of *lumped* states and establish the necessary and sufficient conditions that must be verified by a lumping process to preserve the Markov properties of a chain. We shall study these conditions in greater detail in Chapter III.

In the sequel, we will often refer to the independent subsystems $\mathbf{Q}_I^*$ as being the aggregates of the nearly completely decomposable system $\mathbf{Q}$, to $\mathbf{P}$ as the aggregative matrix of $\mathbf{Q}$, and to $X_I(1)$ and $x_{i_I}(1)$ as the macro- and microaggregation variables or probabilities.

A good survey of the extensive literature on variable aggregation that exists in economics can be found in Green (1964).

1.5 Multilevel Decomposability

1.5.1 The system of variables discussed in the foregoing section can be represented as a two-level hierarchy with the microaggregation variables at the lower level and the macrovariables at the higher level. It is clear, and Simon and Ando (1961) do not fail to observe it, that such a hierarchy may be extended to more than two levels, each variable at a certain level being an aggregate of variables of the immediately lower level. An example of a nearly completely decomposable matrix corresponding to a three-level hierarchy would be

$$\left[\begin{array}{cc|cc} \mathbf{Q}_1 & \mathbf{R}_1 & \mathbf{S}_1 & \mathbf{S}_2 \\ \mathbf{R}_2 & \mathbf{Q}_2 & \mathbf{S}_3 & \mathbf{S}_4 \\ \hline \mathbf{S}_5 & \mathbf{S}_6 & \mathbf{Q}_3 & \mathbf{R}_3 \\ \mathbf{S}_7 & \mathbf{S}_8 & \mathbf{R}_4 & \mathbf{Q}_4 \end{array}\right],$$

where the order of magnitude of the elements of submatrices $\mathbf{Q}$ is larger than that of matrices $\mathbf{R}$, and their order of magnitude being larger than that of matrices $\mathbf{S}$. At the first level of aggregation there could be four aggregative variables, one for each matrix $\mathbf{Q}$; at the second level of aggregation two aggregative variables would correspond to the partitions indicated by the dashed lines.

More generally, we will say that to the $L+1$ levels $l = 0, \ldots, L$ of a nearly completely decomposable hierarchy may correspond L levels of aggregation $l = 1, \ldots, L$, and that such a hierarchy is representable by a *L-level nearly completely decomposable matrix*. In such a system the time instant $T_{2,l}$ beyond which the whole system ceases behaving as a set of nearly independent subsystems at the lth level of aggregation is also the time instant $T_{0,l+1}$ at which subsystems of the adjacent upper level $l+1$ of aggregation start moving toward their internal equilibrium (which they reach at time $T_{1,l+1}$):

$$T_{2,l} \equiv T_{0,l+1}, \qquad l = 1, \ldots, L-1.$$

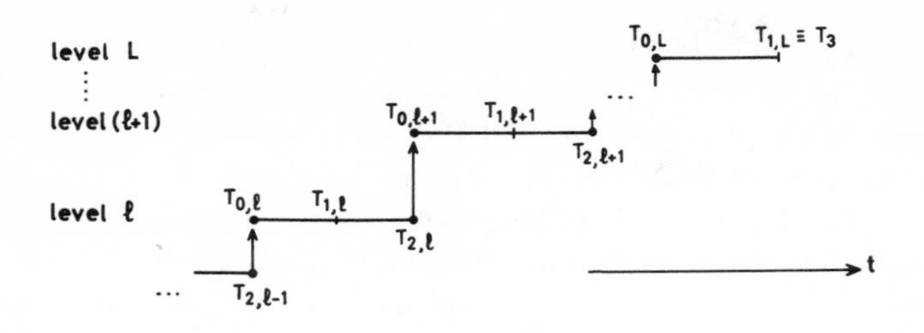

FIGURE 1.1 Time relations in multilevel nearly completely decomposable hierarchies.

The time instant T_3 at which a global equilibrium is reached throughout the whole system is the time instant $T_{1,L}$ at which equilibrium is reached among the aggregative variables in terms of which the whole system is described at the uppermost level L. These time relations among levels are schematized in Fig. 1.1.

Each level l is a level of aggregation provided there exists an ε_l small enough so that

$$T_{1,l-1} < (T_{2,l-1} \equiv T_{0,l}) \qquad \text{for} \quad l = 2, \ldots, L.$$

The analysis of the entire structure then proceeds first from the lowest level of aggregation through each adjacent level up to the highest one; at each level of aggregation the system is viewed as a set of independent aggregates; the short-term equilibrium of each aggregate is analyzed in terms of aggregative variables descriptive of the short-term equilibrium reached by its subaggregates, which were considered at the adjacent lower level. The equilibrium ultimately obtained in this way at the uppermost level defines *in terms of macrovariables* the long-term equilibrium of the whole system. From this long-term equilibrium one can then derive the long-term equilibrium of the aggregates, in terms of microvariables, by proceeding downward successively through each level.

1.5.2 In steady-state analysis, this approach gives rise to the following procedure. For $l = 0, \ldots, L-1$, call $\mathbf{v}^{*[l]}(1_I)$ the steady-state vector of each aggregate $\mathbf{P}_I^{*[l]}$; with these vectors one can construct an aggregative matrix $\mathbf{P}^{[l+1]}$, the elements of which are

$$p_{IJ}^{[l+1]} = \sum_{i \in I} v_{i_I}^{*[l]}(1_I) \sum_{j \in J} p_{i_I j_J}^{[l]}; \tag{1.34}$$

I and J are the sets of entries of aggregates $\mathbf{P}_I^{*[l]}$ and $\mathbf{P}_J^{*[l]}$, respectively, and the $p_{i_I j_J}^{[l]}$ are the elements of a matrix $\mathbf{P}^{[l]}$ that is nearly completely decomposable:

$$\mathbf{P}^{[l]} = \mathbf{P}^{*[l]} + \varepsilon_{l+1} \mathbf{C}^{[l]}; \tag{1.35}$$

ε_{l+1} is the maximum degree of coupling between aggregates of the level of aggregation $(l+1)$. If $\mathbf{P}^{[l+1]}$ is also nearly completely decomposable, a new aggregative matrix $\mathbf{P}^{[l+2]}$ can be constructed in the same way by means of (1.34), and so on up to $\mathbf{P}^{[L]}$. The steady-state eigenvector of $\mathbf{P}^{[L]}$, say $\mathbf{X}^{[L]}$, gives an approximation to the probabilities of being in each of the aggregates of this uppermost level L. This approximation is clearly obtained by calculating successively for $l = 0, \ldots, L-1$ the aggregate steady-state vectors $\mathbf{v}^{*[l]}(1_I)$ and the aggregative matrices $\mathbf{P}^{[l+1]}$.

If we then proceed in the reverse order, we can obtain from $\mathbf{X}^{[l]}$ an approximation of the equilibrium probabilities for each lower level down to the lowest one; more precisely, an approximation $\mathbf{x}^{[l]}$ to the steady-state vector $\mathbf{X}^{[l]}$ of each matrix $\mathbf{P}^{[l]}$, $l = L-1, L-2, \ldots, 0$, can be obtained by means of the recurrent relations

$$x_{i_I}^{[L-1]} = v_{i_I}^{*[L-1]}(1_I)\, X_I^{[L]}, \qquad x_{i_I}^{[l]} = v_{i_I}^{*[l]}(1_I)\, x_{I_J}^{[l+1]}, \quad l = L-2, \ldots, 0, \tag{1.36}$$

where I_J designates a state I in an aggregate J of the upper level.

The vector $\mathbf{x}^{[0]}$ ultimately obtained in this way is an approximation to the steady-state vector of the whole system.

In order to alleviate the notation, we shall from this point on omit specifying the level of aggregation whenever only one level is being considered; in this case, we shall simply use the notation introduced in the preceding sections, and adopt the convention

$$\mathbf{Q} \equiv \mathbf{P}^{[0]}, \qquad \mathbf{P} \equiv \mathbf{P}^{[1]}, \qquad \mathbf{v}^*(1_I) \equiv \mathbf{v}^{*[0]}(1_I),$$
$$\varepsilon \equiv \varepsilon_1, \qquad \mathbf{X}(1) \equiv \mathbf{X}^{[1]}, \qquad \mathbf{x}(1) \equiv \mathbf{x}^{[0]}, \quad \ldots.$$

1.5.3 Finally, one point should not be overlooked. It is only at the first level of aggregation that the maximum degree of coupling $\varepsilon \equiv \varepsilon_1$ is directly expressed (by 1.4) as a maximum row sum of off-diagonal elements of the *original* matrix $\mathbf{Q}$. At all the upper levels $l > 1$, ε_l is a maximum row sum of off-diagonal elements of an *aggregative* matrix $\mathbf{P}^{[l-1]}$. Therefore, it is only at the first level that ε_1 is *equal* to the maximum probability of leaving, upon a transition, any subset of elementary states that corresponds to an aggregate. At all the upper levels $l > 1$, this same maximum probability, which we shall now denote ω_l, is *an upper bound* for ε_l since in this case, and this can be easily verified by resolving the recurrence (1.34), ε_l is a sum of off-diagonal elements of $\mathbf{Q}$ that is averaged by the steady-state vectors of aggregates $\mathbf{P}_I^{*[l-2]}$, $\mathbf{P}_J^{*[l-3]}, \ldots, \mathbf{P}_K^{*[0]}$. This upper bound ω_l can be attained. For example, ε_l will be equal to the maximum probability ω_l of leaving any subset of elementary states that corresponds to an aggregate $\mathbf{P}_I^{*[l-1]}$ if, *in the original matrix* $\mathbf{Q}$, the submatrices that couple an aggregate of level $l-2$ to those which do not belong to the same aggregate of level $l-1$, are each a matrix whose row sums are identical. For example, in the two-level nearly completely decomposable matrix given at the beginning of this section, ε_2 would be equal to the maximum probability of leaving one of the two subsets of elementary states defined by the dashed line partition if row sums are equal in each matrix $[\mathbf{S}_1\ \mathbf{S}_2]$, $[\mathbf{S}_3\ \mathbf{S}_4]$, $[\mathbf{S}_5\ \mathbf{S}_6]$, $[\mathbf{S}_7\ \mathbf{S}_8]$. The reason for this condition can be better understood by considering, for example, the first level of aggregation. We have

$$\sum_{J \neq I, K, \ldots} p_{IJ} = \sum_{J \neq I, K, \ldots} \sum_{i \in I} v_{i_I}^*(1_I) \sum_{j \in J} q_{i_I j_J} = \max_{i_I} \left(\sum_{J \neq I, K, \ldots} \sum_{j \in J} q_{i_I j_J} \right),$$

on condition that $\sum_{J \neq I, K, \ldots} \sum_{j \in J} q_{i_I j_J}$ is the same for all $i \in I$.

The point is that, at the upper levels of aggregation, the maximum degree of coupling between aggregates takes into account the fact that lower-level aggregates are in equilibrium; and the probability of leaving an aggregate by departing from any given state can be conditioned, at these upper levels, by a lower-level equilibrium probability of occupying this state.

1.6 Block Triangular Systems

Ando and Fisher (1963) proved that the Simon–Ando theorems could be extended to the case of *nearly decomposable* matrices, defined by replacing the zeros in a triangular decomposable matrix by small numbers. They showed that, *mutatis mutandis*, the conclusions of Section 1.3 remain true for such systems: the short-term dynamics may be analyzed as if the systems were decomposable, thus ignoring the weak "feedback" between subsystems; the long-term influence of these feedbacks may be analyzed in terms of macro-variables representative of the short-term equilibrium attained by each subsystem. The aggregates that in the Simon–Ando case are the submatrices $\mathbf{Q}_I^*$ are in this case the *submatrices*

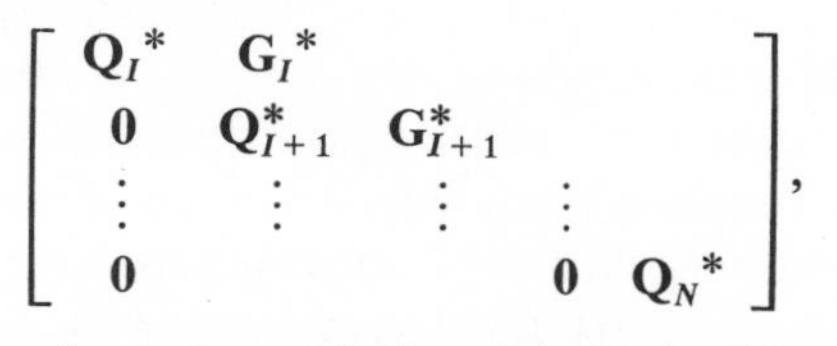

$$\begin{bmatrix} \mathbf{Q}_I^* & \mathbf{G}_I^* & & & \\ \mathbf{0} & \mathbf{Q}_{I+1}^* & \mathbf{G}_{I+1}^* & & \\ \vdots & \vdots & \vdots & \vdots & \\ \mathbf{0} & & & \mathbf{0} & \mathbf{Q}_N^* \end{bmatrix},$$

where the $\mathbf{G}_I^*$ are submatrices of dimensions $(n(I), N - \sum_{J \leqslant I} n(J))$. Such systems are called by Ando and Fisher *nearly decomposable block triangular* systems. We shall not use such models in this work, and the theory and the error analysis developed in the next chapter do not apply as such to them. But the generality of these models is attractive and certainly justifies more investigation.

CHAPTER II

On the Degree of Approximation

We have seen in Chapter I how the Simon–Ando theorems justify the use of a procedure of variable aggregation to approximate the inverse and the steady-state eigenvector of a nearly completely decomposable stochastic matrix. In fact, both the Simon–Ando (1961) and Ando–Fisher (1963) existence theorems aim essentially at proving that whatever standard of accuracy is required from such approximations, there always exists a sufficient degree of near-complete decomposability such that an aggregation procedure would meet this standard. In this chapter, we tackle the other end of the problem. We seek to determine which standard of accuracy may be guaranteed when an aggregation procedure is applied to a given nearly completely decomposable system. In other words, our purpose is to give the investigator a means to estimate beforehand the error to which he is exposed if he uses an aggregation technique. This a priori error estimation is of paramount importance if we want to develop system evaluation methods based on the notion of near-complete decomposability.

Here we are not going to concern ourselves with the approximation of the inverse discussed in Section 1.4.3, but only with the approximation of the eigensystem described in Section 1.4.2; only this latter approximation is useful for all the applications that will be studied later.

The first result of our error analysis will be to show that the process of aggregation described in Section 1.4.2 yields an approximation of accuracy ε, the maximum degree of coupling between aggregates. The second result will be to show that an accuracy of ε^2 is achievable by a procedure that makes use only of the eigencharacteristics of the aggregative matrix and of the steady-state vectors of the aggregates.

Besides, we will also investigate the possibility of approximating by a process of variable aggregation those other eigenvectors of a nearly completely decomposable matrix that are associated with the dominant eigenvalues

$\lambda(1_I)$, $I = 2, \ldots, N$. There is no point in considering an approximation by aggregation of the eigenvectors associated with the other smaller eigenvalues since the first Simon–Ando theorem (Theorem 1.1) shows that the eigenvectors of the completely decomposable matrix $\mathbf{Q}^*$ are direct approximations of the corresponding eigenvectors of the nearly completely decomposable matrix $\mathbf{Q}$.

Finally, special attention will be given to block stochastic matrices because their special properties provide for nearly completely decomposable stochastic matrices a definition of what economists call consistent aggregation. Our results will also cast some light on the relationship between the concept of aggregate and the concept of lumped state in Markov chains; we shall see in the following chapter that this relationship allows aggregates to be conglomerated into classes of equivalence.

2.1 Error Analysis

We shall show in this section that an eigenvector of the original matrix $\mathbf{Q}$ may be approximated by a process of aggregation with an error that tends to zero with ε, the maximum degree of coupling between aggregates, and is a function of the conditioning and of the indecomposability of the aggregates.

2.1.1 First of all, following an approach inspired by the perturbation theory developed by Wilkinson (1965, p. 62 ff), we use the set of eigenvectors of $\mathbf{Q}^*$ as a base to express the set of eigenvectors of $\mathbf{Q}$. To do so, we again restrict ourselves to the case where $\mathbf{Q}^*$, like $\mathbf{Q}$, has a complete set of independent left and right eigenvectors. Then $\mathbf{Q}^*$ is similar to a diagonal matrix, and there exists a nonsingular matrix $\mathbf{H}^*$ such that

$$\mathbf{H}^{*-1}\mathbf{Q}^*\mathbf{H}^* = \mathbf{diag}(\lambda^*(i_I)),$$

where $\mathbf{diag}(\lambda^*(i_I))$ is a diagonal matrix of order n with its (i_I, i_I)th element equal to $\lambda^*(i_I)$. The rows of $\mathbf{H}^{*-1}$ and the columns of $\mathbf{H}^*$ are parallel to the left and right eigenvectors of $\mathbf{Q}^*$, respectively. More precisely, we choose $\tilde{\mathbf{v}}^*(i_I)$ for the i_Ith row of $\mathbf{H}^{*-1}$ and $s^*(i_I)^{-1}\mathbf{v}^*(i_I)$ for the i_Ith column of $\mathbf{H}^*$.

Now let us consider the matrix

$$\mathbf{H}^{*-1}\mathbf{Q}\mathbf{H}^* = \mathbf{H}^{*-1}(\mathbf{Q}^*+\varepsilon\mathbf{C})\mathbf{H}^* = \mathbf{diag}\,(\lambda^*(i_I)) + \varepsilon\mathbf{H}^{*-1}\mathbf{C}\mathbf{H}. \quad (2.1)$$

The eigenvalues are invariant under a similarity transformation, and the left eigenvectors are multiplied by $\mathbf{H}^*$ (see, for instance, Wilkinson, 1965, p. 7). Thus the eigenvalues of $\mathbf{H}^{*-1}\mathbf{Q}\mathbf{H}^*$ are the same as those of $\mathbf{Q}$, and if we denote $\tilde{\boldsymbol{\beta}}(l_L)$ the left[1] eigenvector of $\mathbf{H}^{*-1}\mathbf{Q}\mathbf{H}^*$ associated with $\lambda(l_L)$, for

[1] Similar developments could be made with $\mathbf{H}^{*-1}\mathbf{Q}\mathbf{H}^*$ right eigenvectors; here we consider only the left ones because our main interest lies with the steady-state vector of $\mathbf{Q}$.

$l = 1, \ldots, n(L)$ and $L = 1, \ldots, N$, we have

$$\tilde{\boldsymbol{\beta}}(l_L) = \tilde{\mathbf{v}}(l_L)\mathbf{H}^*, \tag{2.2}$$

$$\tilde{\mathbf{v}}(l_L) = \tilde{\boldsymbol{\beta}}(l_L)\mathbf{H}^{*-1}. \tag{2.3}$$

The vector $\boldsymbol{\beta}(l_L)$ defines the components of $\mathbf{v}(l_L)$ in the space spanned by the complete set of left eigenvectors $\mathbf{v}^*(j_J)$ of the completely decomposable matrix, and since each element $v_{i_I}^*(j_J) = 0$ for all $J \neq I$, it follows from (2.3) that

$$v_{i_I}(l_L) = \sum_{k=1}^{n(I)} \beta_{k_I}(l_L)\, v_{i_I}^*(k_I). \tag{2.4}$$

Let us remark immediately that the elements $\beta_{1_I}(l_L)$ are summations of elements of the corresponding eigenvector of $\mathbf{Q}$ since (2.2) yields

$$\beta_{1_I}(l_L) = s^*(1_I)^{-1} \sum_{i=1}^{n(I)} v_{i_I}(l_L)\, v_{i_I}^*(1_I),$$

which by further substitution of (1.8b) and (1.9b) gives

$$\beta_{1_I}(l_L) = \sum_{i=1}^{n(I)} v_{i_I}(l_L). \tag{2.5}$$

By isolating these elements in (2.4), we have

$$v_{i_I}(l_L) = \beta_{1_I}(l_L)\, v_{i_I}^*(1_I) + \sum_{k=2}^{n(I)} \beta_{k_I}(l_L)\, v_{i_I}^*(k_I). \tag{2.6}$$

We will demonstrate in the next section that the summation on the right-hand side is in ε if $l = 1$. Let us then approximate $v_{i_I}(1_L)$ by a microvariable

$$x_{i_I}(L) = X_I(L)\, v_{i_I}^*(1_I), \tag{2.7}$$

where $X_I(L)$ would be the Ith element of a vector $\mathbf{X}(L)$ of length N. In this case the differences

$$v_{i_I}(1_L) - x_{i_I}(L) = (\beta_{1_I}(1_L) - X_I(L))\, v_{i_I}^*(1_I) + \sum_{k=2}^{n(I)} \beta_{k_I}(1_L)\, v_{i_I}^*(k_I) \tag{2.8}$$

would remain in ε provided that the differences $\beta_{1_I}(1_L) - X_I(L)$ are also in ε. We will show in Section 2.1.3 that this is indeed the case if $\mathbf{X}(L)$ is an eigenvector of the aggregative matrix $\mathbf{P}$ whose elements are defined by (1.25).

Before demonstrating these points, it might be helpful to remark that in steady-state analysis ($L = 1$), Eq. (2.8) may receive an intuitive interpretation. Because of (2.5), $\beta_{1_I}(1_1)$ is the exact steady-state probability of being in any one state i_I of the subsystem $\mathbf{Q}_{II}$, and therefore the difference on the right-hand side of (2.8) accounts for the deviation of the macrovariable $X_I(1)$ from its exact value. The summation on the right-hand side of (2.8) measures the error caused by the use of $v^*_{i_I}(1_I)$ to approximate the marginal probability of being in state i_I, given that it is in subset I. Indeed from (2.6) it follows that

$$\sum_{k=2}^{n(I)} \beta_{k_I}(1_1)\, v^*_{i_I}(k_I) = v_{i_I}(1_1) - \beta_{1_I}(1_1)\, v^*_{i_I}(1_I)$$
$$= \beta_{1_I}(1_1)\{\beta_{1_I}^{-1}(1_1)\, v_{i_I}(1_1) - v^*_{i_I}(1_I)\}, \tag{2.9a}$$

where $\beta_{1_I}^{-1}(1_1)\, v_{i_I}(1_1)$ is this exact marginal steady-state probability of being in state i_I given that it is in subset I.

2.1.2 We shall follow a similar line of argument as Wilkinson (1965, pp. 82–83) to show that the elements $\beta_{k_I}(l_L)$, with $k \neq 1$, in the summation of the right-hand side of (2.6) are in ε provided that $l = 1$.

We introduce the $n \times n$ matrix $\mathbf{\Gamma}$ whose elements are the scalars

$$\gamma_{i_I j_J} = \tilde{\mathbf{v}}^*(i_I)\,\mathbf{C}\,\mathbf{\upsilon}^*(j_J).$$

Since $v^*_{j_J}(i_I) \equiv \upsilon^*_{j_J}(i_I) \equiv 0$ for all i, j, and $I \neq J$, we may simply write

$$\gamma_{i_I j_J} = \tilde{\mathbf{v}}^*(i_I)\,\mathbf{C}_{IJ}\,\mathbf{\upsilon}^*(j_J), \tag{2.9b}$$

where $\mathbf{C}_{IJ}$ is an $n \times n$ matrix whose elements are those of $\mathbf{C}$ at the intersection of the Ith set of rows and the Jth set of columns and zero elsewhere.

The element (i_I, j_J) of the matrix $\mathbf{H}^{*-1}\mathbf{C}\mathbf{H}^*$ occurring in (2.1) is thus equal to $s^*(j_J)^{-1}\gamma_{i_I j_J}$, and since $\lambda(l_L)$ and $\tilde{\boldsymbol{\beta}}(l_L)$ for $l = 1, \ldots, n(L)$, $L = 1, \ldots, N$, are corresponding eigenvalues and eigenvectors of $\mathbf{H}^{*-1}\mathbf{Q}\mathbf{H}^*$, we have

$$\lambda(l_L)\,\tilde{\boldsymbol{\beta}}(l_L) = \tilde{\boldsymbol{\beta}}(l_L)\,\mathbf{H}^{*-1}\mathbf{Q}\mathbf{H}^*.$$

If we replace $\mathbf{H}^{*-1}\mathbf{Q}\mathbf{H}$ by (2.1) and take the i_Ith column on each side, we have

$$\lambda(l_L)\,\beta_{i_I}(l_L) = \lambda^*(i_I)\,\beta_{i_I}(l_L) + \varepsilon s^*(i_I)^{-1} \sum_{J=1}^{N} \sum_{j=1}^{n(J)} \beta_{j_J}(l_L)\,\gamma_{j_J i_I}$$

or

$$(\lambda(l_L) - \lambda^*(i_I))\,\beta_{i_I}(l_L) = \varepsilon s^*(i_I)^{-1} \sum_{J=1}^{N} \sum_{j=1}^{n(J)} \beta_{j_J}(l_L)\,\gamma_{j_J i_I}. \tag{2.10}$$

If we let $\varepsilon \to 0$ in this equation, the right-hand side tends to zero while the left-hand side tends to $(\lambda^*(l_L) - \lambda^*(i_I))\,\beta_{i_I}(l_L)$. Therefore all elements $\beta_{i_I}(l_L)$ must also tend to zero with ε if i_I and l_L are such that $\lambda^*(i_I) \neq \lambda^*(l_L)$. However, as we did not make any assumption on the simplicity of the nonmaximal eigenvalues of the aggregates $\mathbf{Q}_I{}^*$, this inequality can only be asserted either for $i = 1$ and $l \neq 1$, or for $i \neq 1$ and $l = 1$, respectively. In the latter case, if we isolate the elements $\beta_{jJ}(1_L)$, $j \neq 1$, in (2.10) we have

$$\beta_{i_I}(1_L) = \varepsilon s^*(i_I)^{-1}(\lambda(1_L) - \lambda^*(i_I))^{-1}$$
$$\times \left\{ \sum_{J=1}^{N} \beta_{1_J}(1_L)\,\gamma_{1_J i_I} + \sum_{J=1}^{N} \sum_{j=2}^{n(J)} \beta_{j_J}(1_L)\,\gamma_{j_J i_I} \right\},$$

from which, since the second summation on the right is in ε, for $i \neq 1$ and $L = 1, \ldots, N$, it follows that

$$\beta_{i_I}(1_L) = \varepsilon\{s^*(i_I)(\lambda(1_L) - \lambda^*(i_I))\}^{-1} \sum_{J=1}^{N} \beta_{1_J}(1_L)\gamma_{1_J i_I} + O(\varepsilon^2), \quad (2.11)$$

where $O(\varepsilon^2)$ denotes a term approaching zero as ε^2 when $\varepsilon \to 0$.

2.1.3 In order to show that the differences $\beta_{1_I}(1_L) - X_I(L)$ are also in ε, we need the following theorem, the proof of which is deferred to Appendix I:

Theorem 2.1 *For $L = 1, \ldots, N$, the vector*

$$\tilde{\boldsymbol{\beta}}_1(1_L) = [\beta_{1_1}(1_L), \beta_{1_2}(1_L), \ldots, \beta_{1_N}(1_L)]$$

is the solution of the system

$$\tilde{\boldsymbol{\beta}}_1(1_L)[\mathbf{P} - \lambda(1_L)\mathbf{I}_N] = \varepsilon\tilde{\mathbf{K}}(\varepsilon, L), \quad (2.12)$$

where $\mathbf{P}$ *is the* $N \times N$ *aggregative matrix, the elements of which are defined by* (1.25), *and* $\mathbf{K}(\varepsilon, L)$ *is a vector of elements in* ε,

$$k_I(\varepsilon, L) = -n(I) \sum_{J=1}^{N} \sum_{j=2}^{n(J)} \beta_{j_J}(1_L)\gamma_{j_J 1_I}, \qquad I = 1, \ldots, N. \quad (2.13)$$

The elements of $\tilde{\boldsymbol{\beta}}_1(1_L)$ *satisfy*

$$\sum_{I=1}^{N} \beta_{1_I}(1_L) = \begin{cases} 1, & \textit{for} \quad L = 1, \quad (2.14) \\ 0, & \textit{otherwise}. \quad (2.15) \end{cases}$$

Since the $\beta_{j_J}(1_L)$ are in ε for $j \neq 1$ (see Section 2.1.2), the elements $k_I(\varepsilon, L)$ are also in ε, and thus the right-hand side of (2.12) is in ε^2. Moreover, not all the elements $\beta_{1_J}(1_L)$ of the vector $\boldsymbol{\beta}(1_L)$ are in ε; in particular, for $J = L$, from Eq. (2.2) we deduce

$$\beta_{1_L}(1_L) = s^*(1_L)^{-1}\tilde{\mathbf{v}}(1_L)\mathbf{v}^*(1_L),$$

and this is a scalar product that approaches unity as $\varepsilon \to 0$. The determinant of (2.12) must therefore be at least in ε^2 if this system is to have a nontrivial solution,

$$\det[\mathbf{P} - \lambda(1_L)\mathbf{I}_N] = O(\varepsilon^2). \quad (2.16)$$

This determinantal equation differs only by at most ε^2 from the characteristic equation of $\mathbf{P}$, and so do the roots of these two equations. Thus an eigenvalue $\mu(L)$ of $\mathbf{P}$ exists such that $\lambda(1_L) - \mu(L)$ is at most in ε^2. If $\mathbf{X}(L)$ is the eigenvector corresponding to $\mu(L)$, we have

$$\tilde{\mathbf{X}}(L)[\mathbf{P} - \mu(L)\mathbf{I}_N] = 0. \quad (2.17)$$

If we neglect the terms in ε^2, we can replace $\lambda(1_L)$ by $\mu(L)$ in (2.12), and by substraction of (2.17) we obtain

$$[\tilde{\boldsymbol{\beta}}_1(1_L)-\tilde{\mathbf{X}}(L)][\mathbf{P}-\mu(L)\mathbf{I}_N] = \varepsilon\tilde{\mathbf{K}}(\varepsilon, L). \tag{2.18}$$

Since **P** and its eigenvalues differ by at most ε from $\mathbf{I}_N$ and unity, respectively, all elements of $[\mathbf{P}-\mu(L)\mathbf{I}_N]$ are in ε; by simplifying with ε in (2.18), we see that the difference $\beta_1(1_L)-X(L)$, like $\mathbf{K}(\varepsilon, L)$, is also in ε, which is what we had to prove.

The solution yielded by (2.18), where the matrix $[\mathbf{P}-\mu(L)\mathbf{I}_N]$ is singular, is determined by the additional condition that the sum of the differences $\beta_{1_I}(1_L)-X_I(L)$, $I=1,\ldots,N$, be zero. This system will be used in Section 2.4 to estimate these differences with accuracy ε^2.

2.2 Conditioning and Indecomposability of Aggregates

In the preceding section, the elements $\beta_{i_I}(1_L)$, $i \neq 1$, have been taken as the basic components of the differences

$$v_{k_I}(1_L) - x_{k_I}(L) \tag{2.19}$$

in the space spanned by the eigenvectors $\mathbf{v}^*(i_I)$ of the aggregates. The main term of $\beta_{i_I}(1_L)$ is in ε and for $i \neq 1$ equal to

$$\varepsilon \sum_{J=1}^{N} \beta_{1_J}(1_L)\gamma_{1_J i_I}/s^*(i_I)(\lambda(1_L)-\lambda^*(i_I)). \tag{2.20}$$

In this section, we examine how this term depends upon the *conditioning* and the *indecomposability* of the aggregates.

2.2.1 The condition number $s^*(i_I)$, $i \neq 1$, expresses the sensitivity of the eigenvalue $\lambda^*(i_I)$ to perturbations of the aggregate $\mathbf{Q}_I{}^*$, and is also a measure of the rate at which the eigenvalue $\lambda(i_I)$ approaches $\lambda^*(i_I)$ when $\varepsilon \to 0$.

Let us recall that these condition numbers are defined by (1.9a) and obey (1.10), viz.,

$$|s^*(i_I)| = |\tilde{\mathbf{v}}^*(i_I)\mathbf{v}^*(i_I)| \leqslant 1.$$

Since $\mathbf{Q}^*$ has only linear elementary divisors and is similar to a diagonal matrix, we know (see, e.g., Wilkinson, 1965, p. 76) that for a sufficiently small ε there is an eigenvalue $\lambda(i_I)$ of **Q** that is equal to the convergent integer power series

$$\lambda(i_I) = \lambda^*(i_I) + k_1\varepsilon + k_2\varepsilon^2 + \cdots,$$

whatever the multiplicity of $\lambda^*(i_I)$ may be. The coefficient k_1 of the term in ε of this series is given by (Wilkinson, 1965, p. 69)

$$k_1 = \frac{\tilde{\mathbf{v}}^*(i_I)\mathbf{C}\mathbf{v}^*(i_I)}{s^*(i_I)} = \frac{\tilde{\mathbf{v}}^*(i_I)\mathbf{C}_{II}\mathbf{v}^*(i_I)}{s^*(i_I)} = \frac{\gamma_{i_I i_I}}{s^*(i_I)}.$$

The element $\gamma_{i_I i_I}$ is bounded:

$$|\gamma_{i_I i_I}| = \left| \sum_{j=1}^{n(I)} v_j^*(i_I) \sum_{k=1}^{n(I)} v_k^*(i_I) c_{k_I j_I} \right|$$
$$\leqslant \sum_{j=1}^{n(I)} |v_j^*(i_I)| \sum_{k=1}^{n(I)} |v_k^*(i_I)| \, |c_{k_I j_I}| \leqslant 1.$$

Thus, the larger the value of $s^*(i_I)$, the closer to $\lambda^*(i_I)$ the corresponding eigenvalue of $\mathbf{Q}$ is. Equivalently, we may consider $\varepsilon \mathbf{C}_{II}$ as a perturbation of the aggregate $\mathbf{Q}_I^*$; then $s^*(i_I)$ reflects the sensitivity of the eigenvalue $\lambda^*(i_I)$ to this perturbation.

Only one condition number $s^*(i_I)$ is thus involved in the sensitivity of $\lambda^*(i_I)$, so that an aggregate may be well conditioned with respect to one eigenvalue and ill conditioned with respect to another. On the other hand, as shown by Eq. (2.8), all condition numbers $s^*(k_I)$, $k = 2, \ldots, n(I)$, are, as the corresponding $\beta_{k_I}(1_L)$, involved in each difference $v_{i_I}(1_L) - x_{i_I}(L)$ of a given subset I. Hence, only if all $s^*(k_I)$, $k(\neq 1) \in I$, are large could such a difference be small and insensitive to perturbations of $\mathbf{Q}_I^*$.

The $s^*(i_I)$ can be arbitrarily small and hence the $\beta_{i_I}(1_L)$, $i \neq 1$, arbitrarily large. Wilkinson (1965, p. 85), however, remarks that the $s^*(i_I)$ are related by

$$\forall_{i_I}: \quad |s^*(i_I)^{-1}| \leqslant 1 + \sum_{j_J \neq i_I} |s^*(j_J)^{-1}|,$$

which precludes the possibility of only one $s^*(i_I)^{-1}$ being large; it is then likely to have large $s^*(i_I)^{-1}$ almost equal and opposite, so that correlated perturbations in two directions $\mathbf{v}^*(i_I)$ almost cancel each other out.

Wilkinson gives examples of matrices that are ill conditioned with respect to certain eigenvalues; these matrices have condition numbers of the order of 10^{-7} to 10^{-19}. Hermitian matrices, on the other hand, are well conditioned with respect to the eigenvalue problem. These matrices have the property that each right eigenvector has the same elements as the complex conjugate of the corresponding left eigenvector, and so the corresponding condition number is, for the normalization we have chosen, equal to the sum of the squares of these elements and thus always of an order of magnitude near unity.

In any case, it is an advantage of aggregation that the approximation (2.19) depends only on the conditioning of the aggregates and not on the conditioning of the original matrix. Aggregation is therefore well suited to the analysis of large matrices that are ill conditioned with respect to the eigenvector problem, but that are nearly completely decomposable into well-conditioned aggregates. This is not an unlikely situation since, by definition, aggregates regroup elements of approximately the same magnitude and may therefore depart less from the Hermitian property than the original matrix. An extensive discussion on the properties of condition numbers can be found in Wilkinson (1965, p. 85 ff); $s^*(i_I)$ values are given in Appendix III for the aggregates of an example treated there.

2.2.2 The $\beta_{i_I}(1_L)$, $i \neq 1$, decrease in module when the degree of *indecomposability* of the aggregates increases. Indeed, since $\lambda(1_L)$ approaches unity as $\varepsilon \to 0$, the modules of the divisors $\lambda(1_L)-\lambda^*(i_I)$ in (2.20) increase, ε being comparatively small, with the difference

$$\min_{\substack{i \in I \\ i \neq 1}} |1-\lambda^*(i_I)|, \tag{2.21}$$

which can be considered as a measure of the indecomposability of aggregate $\mathbf{Q}_I^*$. Moreover, we have

$$\min_{\substack{i \in I \\ i \neq 1}} |1-\lambda^*(i_I)| \geqslant 1-|\lambda^*(2_I)|, \tag{2.22}$$

the equality being verified if this second largest eigenvalue in module $\lambda^*(2_I)$ is real and positive. Upper bounds for the module of the second largest eigenvalue of nonnegative indecomposable matrices exist. These upper bounds allow lower bounds of aggregate indecomposability to be defined in terms of the aggregate elements. So far the best upper bound for $|\lambda^*(2_I)|$ seems to have been defined by Bauer *et al.* (1969); it is expressed in terms of the aggregate matrix elements and the positive eigenvectors $\mathbf{v}^*(1_I)$ and $\boldsymbol{\upsilon}^*(1_I)$—all elements of the latter being equal to $n(I)^{-1}$—associated with the first eigenvalue $\lambda^*(1_I)=1$:

$$|\lambda^*(2_I)| \leqslant \min\left\{ \max_{1 \leqslant \mu,\rho \leqslant n(I)} \frac{1}{2}\sum_{i=1}^{n(I)} v_{i_I}^*(1_I)\left|\frac{q_{i_I\mu_I}^*}{v_{\mu_I}^*(1_I)}-\frac{q_{i_I\rho_I}^*}{v_{\rho_I}^*(1_I)}\right|, \right.$$
$$\left. \max_{1 \leqslant \mu,\rho \leqslant n(I)} \frac{1}{2}\sum_{i=1}^{n(I)} |q_{\mu_I i_I}^*-q_{\rho_I i_I}^*| \right\}. \tag{2.23}$$

Bauer *et al.* (1969) show in particular that this bound improves upon those previously obtained by Birkhoff (1948), Hopf (1963), and Ostrowski (1964). How good this bound (2.23) is can be judged on the aggregates $\mathbf{Q}_1^*, \mathbf{Q}_2^*, \mathbf{Q}_3^*$ of the example treated in Appendix III.

The bound (2.23) fails, i.e., becomes equal to 1, if there are two rows (and two columns) such that in each column (in each row) there is one element equal to zero. In this case the Lynn–Timlake (1969) bound may be a useful substitute for (2.23). This bound is only valid for nonnegative matrices that are primitive, i.e., have only one eigenvalue of maximum module (in our case equal to 1). This bound is given by

$$|\lambda^*(2_I)| \leqslant \left\{1 - m^\gamma \frac{1-m^{\gamma-1}[1-mn(I)]}{1-m^{\gamma-1}(1-m)}\right\}^{1/\gamma}, \tag{2.24}$$

where $m = \min_{i,j}(q^*_{i_I j_J} \neq 0)$, and where γ is the smallest integer such that all elements of $\mathbf{Q}_I^{*\gamma}$ are strictly positive. Since the aggregates are supposed to be indecomposable, a sufficient condition of primitivity is that all elements of the diagonal be strictly positive (Marcus and Minc, 1964, p. 128); and in this case $\gamma \leqslant n(I)-1$.

Bounds like (2.23) and (2.24) give an indication as to how the aggregates $\mathbf{Q}_I{}^*$ should optimally be constructed from the elements of the original matrix. Each stochastic matrix $\mathbf{Q}_I{}^*$ is obtained by adding to each row i_I of the principal submatrix $\mathbf{Q}_{II}$ of $\mathbf{Q}$ the off-diagonal element sum

$$\sum_{J \neq I} \sum_{j} q_{i_I j_J} \leqslant \varepsilon.$$

So far we have left unspecified the way this sum should be distributed over the elements of $\mathbf{Q}_{II}$. It is now clear that, in order to optimize the approximation, this distribution must minimize the upper bound (2.23) or (2.24). A simple way to do this is to distribute this sum so as to maximize, for each row i_I,

$$m_{i_I} = \min_{j \in I} (q^*_{i_I j_I}).$$

Aggregate indecomposability will turn out in Section 2.5 to be a major limitation in multilevel aggregation. If we evaluate the maximum value of (2.20), we could, in principle, predict the maximum error implied in the approximation of the eigenvectors $\mathbf{v}(1_L)$ by the microvariables $x_{i_I}(L)$, $L = 1, \ldots, N$; the aggregate minimum indecomposability could be assessed by means of the upper bound (2.23) or (2.24), and the values of the condition numbers $s^*(i_I)$ could be obtained from the set of aggregate eigenvectors $\mathbf{v}^*(i_I)$, $\mathfrak{v}^*(i_I)$. In practice, however, this set of eigenvectors may require too much computation. We will see in Section 2.4 that it is possible to estimate the error without relying on the condition numbers. First, we must make a digression and consider the special case of nearly completely decomposable systems that are also block stochastic.

2.3 Block Stochastic Systems

It is interesting to examine the specific case where the nearly completely decomposable matrix is at the same time block stochastic; such matrices are indeed particularly well suited for an analysis by aggregation.

Two cases must be distinguished: a nearly completely decomposable matrix may be, as we shall call it, either *row block stochastic* or *column block stochastic*; in the first case the sum of each row's elements in respective submatrices $\mathbf{Q}_{IJ}$ are equal; in the second case the sums of each column's elements are equal.

2.3.1 ***Row Block Stochastic Matrices*** More precisely, we say that the nearly completely decomposable matrix **Q** is *also row block stochastic* if, for I, $J = 1, \dots, N$, there exist constants W_{IJ} such that

$$\sum_{j=1}^{n(J)} q_{i_I j_J} = W_{IJ}, \qquad i = 1, \dots, n(I), \tag{2.25}$$

and since **Q** is stochastic, such that

$$\sum_{J=1}^{N} W_{IJ} = 1, \qquad I = 1, \dots, N.$$

In this case, from Theorem 2.1 we have

Corollary 2.1 *If a matrix* **Q** *is row block stochastic according to the definition given above, the eigenvalues of the aggregative matrix* **P**, *whose elements are given by* (1.25), *are also eigenvalues of* **Q**, *and the Ith element* ($I = 1, \dots, N$) *of each eigenvector of* **P** *is the sum of the elements in the Ith subset of the corresponding eigenvector of* **Q**.

PROOF If **Q** is row block stochastic, the row sums of each matrix $\mathbf{C}_{IJ}$ defined in Section 2.1.2 are also equal and, for $I \neq J$, identical to $\varepsilon^{-1} W_{IJ}$. Thus

$$\gamma_{i_I 1_J} = \tilde{\mathbf{v}}^*(i_I)\,\mathbf{C}_{IJ}\,\mathbf{v}^*(1_J) = n(J)^{-1}\varepsilon^{-1} W_{IJ} \sum_{k=1}^{n(I)} v_{k_I}^*(i_I).$$

For all I, J we have therefore $\gamma_{i_I 1_J} = 0$, if $i \neq 1$, since $\tilde{\mathbf{v}}^*(i_I)$, $i \neq 1$, is orthogonal to any vector that, like $\mathbf{v}^*(1_I)$, has identical elements. Thus, for $i \neq 1$, $J, L = 1, \dots, N$, we have

$$k_J(\varepsilon, L) = 0$$

from which by Theorem 2.1 it follows that ($L = 1, \dots, N$)

$$\tilde{\boldsymbol{\beta}}_1(1_L)[\mathbf{P} - \lambda(1_L)\mathbf{I}_N] = 0. \tag{2.26}$$

This completes the proof since it implies that the vectors $\tilde{\boldsymbol{\beta}}_1(1_L)$ whose elements are given by (2.5) are the left eigenvectors of **P**, and $\lambda(1_L)$ are its eigenvalues. An identical approach may evidently be followed to prove the same result for the right eigenvectors. Neither the above proof, nor the proof of Theorem 2.1 which is involved require ε to be small. This explains why the corollary does not specify that Q must be nearly completely decomposable.

Let us note that, since the aggregative matrix **P** is identical to the matrix $[W_{IJ}]$ when **Q** is row block stochastic, the first part of this corollary is similar to Haynsworth's (1959) theorem, which states that the eigenvalues of $[W_{IJ}]$ are also eigenvalues of **Q**.

The significance of Corollary 2.1 is that nearly completely decomposable matrices that are row block stochastic are well suited for an analysis by aggregation. The macrovariables $X_I(L)$ are exact summations over elements of the corresponding eigenvector $\mathbf{v}(1_L)$ of the original matrix; in particular, the macrovariables $X_I(1)$ yield, in steady-state analysis, exact values for the long-term probabilities of being in any one state of subset I. The residual errors [see the differences (2.8)]

$$\sum_{k=2}^{n(I)} \beta_{k_I}(1_L)\, v_{i_I}^*(k_I)$$

are only due to the fact that, in the aggregation procedure, the stochastic submatrices $\mathbf{Q}_I{}^*$ take the place of the submatrices $\mathbf{Q}_{II}$ that are only sub-stochastic matrices with row sums W_{II} not equal to unity:

$$1 > W_{II} = 1 - \sum_{J \neq I} W_{IJ} \geqslant 1 - \varepsilon. \tag{2.27}$$

Finally it is noteworthy that this block stochastic condition for nearly completely decomposable systems is the same as the condition for *acceptable* or *consistent* aggregation in input–output economic analysis [see, for example, Theorem 1 by Ara (1959) and Theil (1957, pp. 118–119)].

2.3.2 ***Column Block Stochastic Systems*** Let us now see how the aggregation procedure works when the *column sums* instead of the row sums are identical in the submatrices $\mathbf{Q}_{IJ}$ of a nearly completely decomposable matrix $\mathbf{Q}$. This means that there exist constants V_{IJ} such that, for any $I, J = 1, \ldots, N$,

$$\sum_{i=1}^{n(I)} q_{i_I j_J} = V_{IJ} \qquad \text{for all} \quad j = 1, \ldots, n(J). \tag{2.28}$$

The aggregates $\mathbf{Q}_I{}^*$ of such nearly completely decomposable matrices can always be constructed so that they are *doubly stochastic matrices*, i.e., so that both their row sums and their column sums are equal to unity: the off-diagonal element row sums

$$\sum_{J \neq I} \sum_{j=1}^{n(I)} q_{i_I j_J}, \qquad i = 1, \ldots, n(I),$$

need simply to be evenly shared out among the $n(I)$ columns of $\mathbf{Q}_{II}$. Then each column sum in $\mathbf{Q}_I{}^*$ will be equal to

$$n(I)^{-1} \sum_{J=1}^{N} n(J) V_{IJ} = 1 \tag{2.29}$$

since the row sums of $\mathbf{Q}$ are equal to 1.

Doubly stochastic matrices have the property that all elements of their steady-state eigenvectors are equal, i.e., all equilibrium states are equally probable. Thus, for all aggregates $\mathbf{Q}_I{}^*$ of a nearly completely decomposable stochastic matrix that is also column block stochastic, we have

$$v_{i_I}^*(1_I) = n(I)^{-1}, \qquad i = 1, \ldots, n(I), \quad I = 1, \ldots, N. \tag{2.30}$$

The aggregative procedure becomes therefore rather straightforward. The microvariables $x_{i_I}(L)$ that approximate the eigenvectors $\mathbf{v}(1_L)$, $L = 1, \ldots, N$, of $\mathbf{Q}$ are given by [Eq. (2.7)]

$$x_{i_I}(L) = X_I(L)\, v_{i_I}^*(1_I),$$

which reduces here to

$$x_{i_I}(L) = X_I(L)\, n(I)^{-1},$$

where $\mathbf{X}(L)$ is eigenvector of the aggregative transition matrix $\mathbf{P}$. The elements of this matrix reduce to

$$p_{IJ} = \sum_{i \in I} n(I)^{-1} \sum_{j \in J} q_{i_I j_J} = V_{IJ} \frac{n(J)}{n(I)}.$$

Moreover, column block stochastic matrices enjoy the following property:

Corollary 2.2 *Aggregation yields the eigenvectors* $\mathbf{v}(1_L)$, $L = 1, \ldots, N$, *of a stochastic nearly completely decomposable matrix* $\mathbf{Q}$ *that is also column block stochastic with an accuracy* ε^2.

PROOF We first prove that all elements $\gamma_{1_I j_J}$ defined by (2.9b) are zero. From (2.9b) and (2.30) it follows that

$$\gamma_{1_I j_J} = n(I)^{-1} \sum_{i \in I} \sum_{k \in J} c_{i_I k_J} \upsilon_{k_J}^*(j_J),$$

and by (2.28) we have

$$\gamma_{1_I j_J} = n(I)^{-1} \varepsilon^{-1} V_{IJ} \sum_{k \in J} \upsilon_{k_J}^*(j_J).$$

Thus $\gamma_{1_I j_J}$ is zero since the eigenvector $\boldsymbol{\upsilon}^*(j_J)$, $j \neq 1$, is orthogonal to any vector that, like $\mathbf{v}^*(1_J)$, has identical nonzero elements. Then it results from Eq. (2.11) that all elements $\beta_{i_I}(1_L)$ are in ε^2, as well as the elements $k_I(\varepsilon, L)$ defined by Eq. (2.13). From Eqs. (2.18) and (2.8) it then follows that the differences $v_{i_I}(1_L) - x_{i_I}(L)$ are also in ε^2, which completes the proof.

Hence, the approximation by aggregation is not only more straightforward but also better in the case of nearly completely decomposable matrices that are column block stochastic than in the case of row block stochastic matrices. As shown by the above proofs, this is due to the fact that aggregates of column block stochastic matrices can be made doubly stochastic. The properties that have been established for row block stochastic matrices will nevertheless be the more instrumental in our later developments; the reason is that row sums like W_{IJ} are transition probabilities between aggregates, which the column sums V_{IJ} are not.

2.3.3 ***Remark*** These properties of row block stochastic matrices may be useful from a measurement point of view. In a nearly completely decomposable system, transitions between individual states of distinct aggregates may be so rare that their probability of occurrence may be difficult to estimate with sufficient accuracy; to attain a sufficient degree of statistical confidence, too many or too sharp observations may have to be made. In this case, it may be advantageous to "row block stochasticize the system," i.e., to measure and to base the analysis on only average transition probabilities $\overline{W}_{IJ}$ between aggregates. The eigenvectors of the aggregative matrix $[\overline{W}_{IJ}]$ are approximations that may be as good, if not better, than the vectors $\mathbf{X}[L]$ of the aggregative matrix that would have been calculated from risky estimations of the transition probabilities $q_{i_I j_J}$ between all pairs of elementary states.

2.4 Error Estimation

Our error analysis of Section 2.1 shows that the classical process of aggregation outlined in Section 1.4 yields an approximation of accuracy ε to any eigenvector of the original matrix associated to a dominant eigenvalue. We show hereafter how an estimate in ε^2 of this approximation can be obtained. The method proposed makes use of only the aggregates $\mathbf{Q}_I^*$, their steady-state vectors, and the eigencharacteristics of the aggregative matrix, the values of which are made available by the process of aggregation itself.

2.4.1 We start by indicating how to obtain an evaluation in ε^2 of the summations that appear in the difference (2.8), and that for convenience we denote $b_k(I, L)$:

$$b_k(I, L) = \sum_{i=2}^{n(I)} \beta_{i_I}(1_L)\, v_{k_I}^*(i_I), \qquad I, L = 1, \ldots, N; \quad k = 1, \ldots, n(I).$$

Since for $i \neq 1$, we have $\sum_{k \in I} v_{k_L}^*(i_I) = 0$, we have also

$$\sum_{k \in I} b_k(I, L) = 0. \tag{2.31}$$

With $\mathbf{B}(I, L)$ denoting the vector of size $n(I)$ with elements $b_k(I, L)$, $k = 1, \ldots, n(I)$, in Appendix II we prove:

Theorem 2.2 *Each vector* $\mathbf{B}(I, L)$, $I, L = 1, \ldots, N$, *defined above is a solution of the* $n(I) \times n(I)$ *system*

$$\tilde{\mathbf{B}}(I, L)[\lambda(1_L)\mathbf{I}_{n(I)} - \mathbf{Q}_I^*] = \tilde{\mathbf{A}}(I, L), \tag{2.32}$$

where $\mathbf{A}(I, L)$ *is a vector of elements*

$$a_k(I, L) = \varepsilon \sum_{J=1}^{N} \beta_{1_J}(1_L) \sum_{j \in J} v^*_{j_J}(1_J)\left(c_{j_J k_I} - v^*_{k_I}(1_I) \sum_{i \in I} c_{j_J i_I}\right) + O(\varepsilon^2) \tag{2.33}$$

for $k = 1, \ldots, n(I)$. *Moreover, for* $L = 1$, *these elements reduce to*

$$a_k(I, 1) = \varepsilon \sum_{J=1}^{N} \sum_{j \in J} x_{j_J}(1)\, c_{j_J k_I} + O(\varepsilon^2), \tag{2.34}$$

where the $x_{j_J}(1)$ *are the steady-state microprobabilities obtained by aggregation and defined by* (2.7).

Let us remark immediately that the system of equations (2.31), (2.32) defines with accuracy ε^2 a unique solution for the vector $\mathbf{B}(I, L)$. Indeed, with Eq. (2.33) one can easily verify that $\sum_{k \in I} a_k(I, L) = O(\varepsilon^2)$; thus, if we neglect $O(\varepsilon^2)$, the matrix of the system (2.31), (2.32) and the augmented matrix of this system have the same rank, and the system is compatible. For $L = 1$, the matrix in (2.32) is singular; but $\mathbf{Q}_I{}^*$ being irreducible, this matrix is of rank $n(I) - 1$, and the system (2.31), (2.32) also yields in this case a unique solution.

The consequences of Theorem 2.2 are clear. The $n(I) \times n(I)$ system (2.32), (2.33)—or (2.32), (2.34), if $L = 1$—to be solved with accuracy ε^2 requires the sheer knowledge of the eigencharacteristics of the aggregative matrix $\mathbf{P}$ since, at their worst, the differences $\lambda(1_L) - \mu(L)$ are in ε^2 and the differences $\beta_I(1_L) - X_I(L)$ are in ε. Thus with these eigencharacteristics, the aggregate steady-state vectors $\mathbf{v}^*(1_J)$ and the elements of matrix $\mathbf{C}$, the system (2.32), (2.33) or (2.32), (2.34) provides an estimate in ε^2 of the summations $b_k(I, L)$. These summations, if the original matrix is row block stochastic, are equal to the differences $v_{k_I}(1_L) - x_{k_I}(L)$.

If the original matrix is not row block stochastic, an estimate of the macrodifferences $\beta_I(1_L) - X_I(L)$ is needed in addition to the estimate of $b_k(I, L)$ [see Eq. (2.8)]. This macrodifference may be evaluated by means of the $N \times N$ system (2.18), because the elements of the vector $\mathbf{K}(\varepsilon, L)$ are themselves functions of the elements $b_k(I, L)$. Indeed, if we replace $\gamma_{j_J 1_I}$ in (2.13) by

$$\gamma_{j_J 1_I} = n(I)^{-1} \sum_{l \in J} \sum_{i \in I} v^*_{l_J}(j_J)\, c_{l_J i_I},$$

which is derived from (2.9b) for the elements of $\mathbf{K}(\varepsilon, L)$, we obtain

$$\begin{aligned} k_I(\varepsilon, L) &= -\sum_{J=1}^{N} \sum_{l \in J} \sum_{j=2}^{n(J)} \beta_{j_J}(1_L)\, v^*_{l_J}(j_J) \sum_{i \in I} c_{l_J i_I} \\ &= -\sum_{J=1}^{N} \sum_{l \in J} b_l(J, L) \sum_{i \in I} c_{l_J i_I}. \end{aligned}$$

To sum up, the refined procedure yielding an ε^2 estimate of the differences $v_{i_I}(1_L)-x_{i_I}(L)$, $i=1,\ldots,n(I)$, consists of the successive evaluation of (a) the vectors $\mathbf{B}(J,I)$, $J=1,\ldots,N$; (b) the vector $\mathbf{K}(\varepsilon,L)$, and (c) the macrodifferences $\beta_{1_I}(1_L)-X_I(L)$, which must be added to the element $b_i(I,L)$ in Eq. (2.8). If the original matrix is row block stochastic, only the vector $\mathbf{B}(I,L)$ is needed.

This procedure preserves the main advantage of aggregation, i.e., only the resolution of small independent subsystems of sizes $n(I)$ and N is required. A numerical example illustrating this procedure is given in Appendix III.

It has also been shown (Courtois and Louchard, 1976) that the approach followed here can be generalized; this generalization yields iteratively the next terms in $\varepsilon^2, \varepsilon^3, \ldots$ of the expansion in ε of the eigencharacteristics of a nearly completely decomposable matrix, aggregation being used at each step of the iterative process.

2.5 Multilevel Aggregation

The system of variables we have studied so far can be viewed as a two-level hierarchy with the aggregative variables $X_I(L)$ at the higher level and the microvariables $x_{i_I}(L)$ at the lower one. We have already indicated in Section 1.5 how such a hierarchy may be extended to more than two levels, each variable at a certain level being an aggregate of variables of the next level down.

In this section, we examine the order of magnitude and the propagation of the error in such a hierarchy. In order to avoid unnecessary notational complexity, we shall limit the discussion to an example of steady-state analysis.

2.5.1 Let us suppose that the nearly completely decomposable matrix $\mathbf{Q}$ is of the form

$$\mathbf{Q} = \left[\begin{array}{cc|ccc} \mathbf{Q}_1 & \mathbf{R}_1 & \mathbf{S}_1 & \mathbf{S}_2 & \mathbf{S}_3 \\ \mathbf{R}_2 & \mathbf{Q}_2 & \mathbf{S}_4 & \mathbf{S}_5 & \mathbf{S}_6 \\ \hline \mathbf{S}_7 & \mathbf{S}_8 & \mathbf{Q}_3 & \mathbf{R}_3 & \mathbf{R}_4 \\ \mathbf{S}_9 & \mathbf{S}_{10} & \mathbf{R}_5 & \mathbf{Q}_4 & \mathbf{R}_6 \\ \mathbf{S}_{11} & \mathbf{S}_{12} & \mathbf{R}_7 & \mathbf{R}_8 & \mathbf{Q}_5 \end{array}\right]. \tag{2.35}$$

The diagonal submatrices $\mathbf{Q}_I$ are square and the order of magnitude of their elements is larger than that of submatrices $\mathbf{R}_I$, which is larger than the order of magnitude of elements in matrices $\mathbf{S}_I$.

We have in this case two levels of aggregation. At the first level, the aggregates $\mathbf{Q}_I{}^*$ are those which can be determined from the subsystems $\mathbf{Q}_I$; at the second level, the aggregates are those which are derivable from the subsystems isolated by dashed lines.

The maximum degree of coupling between aggregates $\mathbf{Q}_I{}^*$ of the first level is ε_1 and is equal to the maximum row sum of the off-diagonal elements of $\mathbf{Q}$ that do not belong to submatrices $\mathbf{Q}_I$. The first-level aggregative matrix of transition probabilities between aggregates $\mathbf{Q}_I{}^*$ is the 5×5 matrix $\mathbf{P}^{[1]}$, the elements $p_{IJ}^{[1]}$ of which are defined by (1.34). Let us define on $\mathbf{P}^{[1]}$ the quantities η_1 and η_2:

$$\eta_1 = \max\{p_{12}^{[1]}, p_{21}^{[1]}, (p_{34}^{[1]}+p_{35}^{[1]}), (p_{43}^{[1]}+p_{45}^{[1]}), (p_{53}^{[1]}+p_{54}^{[1]})\}, \tag{2.36}$$

$$\eta_2 = \max\left\{\left\| \begin{matrix} p_{13}^{[1]} & p_{14}^{[1]} & p_{15}^{[1]} \\ p_{23}^{[1]} & p_{24}^{[1]} & p_{25}^{[1]} \end{matrix} \right\|_\infty, \quad \left\| \begin{matrix} p_{31}^{[1]} & p_{32}^{[1]} \\ p_{41}^{[1]} & p_{42}^{[1]} \\ p_{51}^{[1]} & p_{52}^{[1]} \end{matrix} \right\|_\infty \right\}, \tag{2.37}$$

where $\|\mathbf{A}\|_\infty$ is used to denote for a matrix $\mathbf{A}$ the subordinate norm that is equal to $\max_i(\sum_j |a_{ij}|)$. In plain words, η_1 can be regarded as *the maximum degree of coupling between the aggregates of level* 1 *that belong to the same aggregate of level* 2, *while* η_2 *is clearly identical to* ε_2, *the maximum degree of coupling between aggregates of level* 2.

A consequence of the definition of η_1 and η_2 is, since $\mathbf{P}^{[1]}$ is a stochastic matrix, that

$$p_{II}^{[1]} \geqslant 1-(\eta_1+\eta_2), \qquad I = 1, \ldots, 5. \tag{2.38}$$

Another consequence of the definition of η_1 and η_2 is that, if we assume

$$\varepsilon_2 \equiv \eta_2 \ll \eta_1,$$

then the aggregative matrix $\mathbf{P}^{[1]}$ is itself nearly completely decomposable, and we have

$$\mathbf{P}^{[1]} = \mathbf{P}^{*[1]} + \varepsilon_2 \mathbf{C}^{[1]} = \begin{bmatrix} \mathbf{P}_1^{*[1]} & 0 \\ 0 & \mathbf{P}_2^{*[1]} \end{bmatrix} + \varepsilon_2 \mathbf{C}^{[1]},$$

where

$$\mathbf{P}_1^{*[1]} = \begin{bmatrix} p_{11}^{*[1]} & p_{12}^{*[1]} \\ p_{21}^{*[1]} & p_{22}^{*[1]} \end{bmatrix}, \qquad \mathbf{P}_2^{*[1]} = \begin{bmatrix} p_{33}^{*[1]} & p_{34}^{*[1]} & p_{35}^{*[1]} \\ p_{43}^{*[1]} & p_{44}^{*[1]} & p_{45}^{*[1]} \\ p_{53}^{*[1]} & p_{54}^{*[1]} & p_{55}^{*[1]} \end{bmatrix},$$

and where the elements of $\mathbf{C}^{[1]}$ obey the same relations as those obeyed by matrix $\mathbf{C}$ in Section 1.1.1. If $\mathbf{P}^{[1]}$ is nearly completely decomposable, its steady-state vector $\mathbf{X}^{[1]}$ can be approximated by a vector $\mathbf{x}^{[1]}$ obtained by applying again on $\mathbf{P}^{[1]}$ an aggregation technique; the elements of $\mathbf{x}^{[1]}$ are defined by Eq. (1.36):

$$x_{k_K}^{[1]} = v_{k_K}^{*[1]}(1_K)\, X_K^{[2]}, \qquad k \in K, \quad K = 1, 2,$$

where $\mathbf{v}^{*[1]}(1_K)$ is the steady-state vector of aggregate $\mathbf{P}_K^{*[1]}$, and where $\mathbf{X}^{[2]}$ is the steady-state vector of the 2×2 aggregative matrix $\mathbf{P}^{[2]}$ whose elements are a function of $\mathbf{P}^{[1]}$ elements and of $v^{*[1]}(1_K)$ elements, and are defined by (1.34).

2.5.2 We must first examine the difference between $\mathbf{X}^{[1]}$ and its approximation $\mathbf{x}^{[1]}$. This difference can be evaluated as follows. Since the matrices $\mathbf{P}_K^{*[1]}$ are stochastic, from (2.38) it follows that

$$p_{II}^{*[1]} \geqslant p_{II}^{[1]} \geqslant 1-(\eta_1+\varepsilon_2), \qquad I = 1,\ldots,5, \tag{2.39}$$

and also

$$\sum_{J\neq I} p_{IJ}^{*[1]} \leqslant \eta_1+\varepsilon_2. \tag{2.40}$$

The eigenvalues of these matrices $\mathbf{P}_K^{*[1]}$, say $\lambda^{*[1]}(k_K)$, lie in the closed region of the $\lambda^{*[1]}$-plane, which consists of all the Gerschgorin's disks of centers $p_{II}^{*[1]}$ and radii $\sum_{J\neq I} p_{IJ}^{*[1]}$. Thus we have

$$|p_{II}^{*[1]}-\lambda^{*[1]}(k_K)| \leqslant \sum_{J\neq I} p_{IJ}^{*[1]}, \qquad I,J = 1,\ldots,5, \quad k\in K, \quad K = 1,2. \tag{2.41}$$

Let us now assume that $(\eta_1+\varepsilon_2) < \frac{1}{2}$. This is not so important a restriction; ε_1 is of the same order of magnitude as $(\eta_1+\varepsilon_2)$, and aggregates at the first level would lose their significance if their maximum degree of coupling, and thus the probability of leaving them, could be as large as $\frac{1}{2}$.

If $(\eta_1+\varepsilon_2) < \frac{1}{2}$, relations (2.39) and (2.40) indicate that the matrix $\mathbf{P}^{*[1]}$ is diagonally dominant; and it follows from (2.41) that all eigenvalues of $\mathbf{P}^{*[1]}$ have a positive real part ($\mathbf{P}^{*[1]}$ is positive definite). We can therefore write

$$\min_{k_K} |\lambda^{*[1]}(k_K)| \geqslant p_{II}^{*[1]} - \sum_{J\neq I} p_{IJ}^{*[1]}$$

or

$$p_{II}^{*[1]} - \min_{k_K} |\lambda^{*[1]}(k_K)| \leqslant \sum_{J\neq I} p_{IJ}^{*[1]},$$

or, because of (2.39) and (2.40),

$$1-\min_{k_K} |\lambda^{*[1]}(k_K)| \leqslant 2(\eta_1+\varepsilon_2). \tag{2.42}$$

This inequality (2.42) provides us with an *upper bound* for the maximum indecomposability $1-\min_{k_K}|\lambda^{*[1]}(k_K)|$ of the aggregates $\mathbf{P}_K^{*[1]}$. We know that this indecomposability determines the accuracy we can expect from the approximation of the vector $\mathbf{X}^{[1]}$ by $\mathbf{x}^{[1]}$. In fact, from equation similar to (2.11) that we could establish for the matrix $\mathbf{P}^{[1]}$, it follows that the components of the vector $[\mathbf{X}^{[1]}-\mathbf{x}^{[1]}]$, in the space spanned by the eigenvectors of the aggregates $\mathbf{P}_K^{*[1]}$, have at best an order of magnitude equal to

$$\frac{\varepsilon_2}{1-\min|\lambda^{*[1]}(k_K)|} \geqslant \frac{\varepsilon_2}{2(\eta_1+\varepsilon_2)}.$$

Thus, the smaller η_1, the worse the approximation of $\mathbf{X}^{[1]}$ by $\mathbf{x}^{[1]}$ is likely to be.

Now, $\mathbf{X}^{[1]}$ is itself an approximation to the exact steady-state probabilities $\beta_{1_I}(1_1)$ of being in each subsystem $\mathbf{Q}_{II}$. The order of magnitude of this approximation is ε_1. So the order of magnitude of the differences $\beta_{1_I}(1_1)-x_I^{[1]}$ is at best

$$\varepsilon_1 \pm \varepsilon_2/2(\eta_1+\varepsilon_2),$$

which, since ε_2 is small compared with η_1, is approximately equal to

$$\varepsilon_1 \pm \varepsilon_2/2\eta_1. \tag{2.43}$$

The sign $\pm$ indicates that for any given I, the difference $x_I^{[1]}-X_I^{[1]}$ may add to or may compensate for the difference $X_I^{[1]}-\beta_{1_I}(1_1)$. This value (2.43) is also the order of magnitude of the microdifferences

$$v_{i_I}(1_1) - x_{i_I}^{[0]}, \tag{2.44}$$

where $\mathbf{v}(1_1)$ is the steady-state vector of the whole matrix (2.35), and $[x_{i_I}^{[0]}]$ is its approximation defined by (1.37), i.e.,

$$x_{i_I}^{[0]} = v_{i_I}^*(1_I)x_I^{[1]},$$

where $\mathbf{v}^*(1_I)$ is the steady-state vector of aggregate $\mathbf{Q}_I{}^*$. Indeed, the differences (2.44) obey relations similar to (2.8), i.e.,

$$v_{i_I}(1_1) - x_{i_I}^{[0]} = (\beta_{1_1}(1_I)-X_I^{[1]})\, v_{i_I}^*(1_I) + \sum_{r=2}^{n(I)} \beta_{r_I}(1_1)\, v_{i_I}^*(r_I), \tag{2.45}$$

where, on the right-hand side, the difference has an order of magnitude equal to (2.43), while the summation, which is independent of the upper level aggregation, remains of order ε_1.

As, in general, $\eta_1+\varepsilon_2 \ll 1$, expression (2.43) shows how dependent on the indecomposability of the aggregates of the first level is an approximation obtained by a two-level aggregation technique.

2.5.3 This reasoning carried out on a two-level nearly completely decomposable matrix can be easily generalized to an arbitrary number of levels.

We define η_l, $l=0,\dots,L-1$, as *the maximum degree of coupling between the aggregates of level l that belong to a same aggregate of level* $(l+1)$. If $\varepsilon_{l+1} \ll \eta_l$, the aggregative matrix $\mathbf{P}^{[l]}$ of macrotransition probabilities between the aggregates of level l will be nearly completely decomposable. Then the degree of approximation that must be ascribed to the microvalues $x_{i_I}^{[0]}$ obtained by successive aggregation of the matrices $\mathbf{P}^{[1]}, \mathbf{P}^{[2]}, \dots, \mathbf{P}^{[L-1]}$, is given by

$$\varepsilon_1 + \sum_{k=2}^{L} \pm \frac{\varepsilon_k}{2\eta_{k-1}}. \tag{2.46}$$

Indeed, to the error made on $\mathbf{X}^{[1]}$, the error made on $\mathbf{X}^{[2]}, \mathbf{X}^{[3]}, \dots, \mathbf{X}^{[L-1]}$, may have to be added.

2.5.4 This degree of approximation (2.46) is reduced when some of (or all) the matrices $\mathbf{P}^{[l]}$ are row block stochastic; in this case only the last summation in (2.45) must be taken into consideration at the corresponding levels of aggregation. The matrix defined by (2.35), for example, would be block stochastic, as well as its first level aggregation matrix $\mathbf{P}^{[1]}$, if the row sums were equal in each submatrix

$$\mathbf{R}_I, \quad \mathbf{S}_I, \quad \text{and} \quad \begin{bmatrix} \mathbf{S}_1 & \mathbf{S}_2 & \mathbf{S}_3 \\ \mathbf{S}_4 & \mathbf{S}_5 & \mathbf{S}_6 \end{bmatrix} \begin{bmatrix} \mathbf{S}_7 & \mathbf{S}_8 \\ \mathbf{S}_9 & \mathbf{S}_{10} \\ \mathbf{S}_{11} & \mathbf{S}_{12} \end{bmatrix}.$$

A numerical example of a two-level aggregation is given at the end of Appendix III.

2.5.5 The lesson to be drawn from this section is simply the following. For an approximation by multilevel aggregation to be satisfactorily accurate, the couplings ε_l between aggregates must not only be small at each level, but they must also be small compared with the intraaggregate couplings η_{l-1}. This is clearly in agreement with our first-level error analysis, which showed that the approximation depends not only on ε_1 but also on the indecomposability of the lowest level aggregates. In both cases we fall back upon the premises of near-complete decomposability: interactions *between* aggregates must be small compared with interactions *within* aggregates—"at any level" let us simply add.

2.6 A Posteriori Error Bound

We close this chapter by what is still only a half-baked idea that needs to be worked on.

It is now clear that the approximation made by aggregation is primarily dependent on the maximum degree of coupling between aggregates. This maximum degree of coupling has been defined as the maximum probability of leaving any aggregate by departing from any state of this aggregate, *given that this state is occupied.*

Intuitively, one is inclined to think that the error made by aggregation is in reality proportional to the maximum unconditional probability of leaving an aggregate; this unconditional probability is the probability of leaving an aggregate from any given state multiplied by the probability of occupying this state. In steady-state the maximum of this probability is given by

$$\max_{i_I} \left(\frac{v_{i_I}(1_1)}{\sum_{i \in I} v_{i_I}(1_1)} \sum_{J \neq I} \sum_{j \in J} q_{i_I j_J} \right). \tag{2.47}$$

We have, of course, no way of calculating this unconditional probability, since by hypothesis the exact steady-state vector $\mathbf{v}(1_1)$ is unknown; but an approximation of this probability (2.47), let us call it ε_{eff}, is available:

$$\varepsilon_{\text{eff}} = \max_{i_I} \left(v_{i_I}^*(1_I) \sum_{J \neq I} \sum_{j \in I} q_{i_I j_J} \right).$$

ε_{eff} is an approximation in $O(\varepsilon^2)$ of the probability (2.47), since $v_{i_I}^*(1_I)$ is an approximation in $O(\varepsilon)$ of $v_{i_I}(1_1)/\sum_{i \in I} v_{i_I}(1_1)$, and the summation $\sum_{J \neq I} \sum_{j \in J} q_{i_I j_J}$ is also in $O(\varepsilon)$.

We have also

$$\varepsilon_{\text{eff}} \leqslant \varepsilon \max_{i_I} v_{i_I}^*(1_I) < \varepsilon,$$

since the aggregates are indecomposable, and thus $v_{i_I}^*(1_I) < 1$ for any i_I. This sort of "effective" maximum degree of coupling ε_{eff} is likely to be a more accurate estimate of the order of magnitude of the approximation made by aggregation than ε; ε_{eff} can be evaluated once the aggregate steady-state vectors $\mathbf{v}^*(1_I)$ are known.

For example, in the numerical problem treated in Appendix III one can easily verify in Table A3.1 that the maximum error occurs for the state (1_1) and is given by

$$v_{1_1}(1_1) - x_{1_1}(1) = 0.25 \times 10^{-3} = 0.25\varepsilon.$$

The upper bound of this error is

$$\varepsilon_{\text{eff}} = v_{2_2}^*(1_2)\varepsilon = 0.42\varepsilon.$$

2.7 Conclusions

To sum up this chapter, let us say that the Simon–Ando approximation is of the same order of magnitude as the maximum degree of coupling ε between aggregates, provided that these aggregates are sufficiently well conditioned and indecomposable. This latter restriction is a major limitation for higher order aggregation since in this case aggregates at an upper level of aggregation are nearly decomposable into those of the adjacent lower level.

An accuracy of ε^2 may be achieved by aggregation in the evaluation of eigenvectors associated with dominant eigenvalues. The procedure essentially uses the eigencharacteristics of the aggregative matrix and the steady-state vectors of the aggregates.

Nearly completely decomposable matrices that are also block stochastic have, from the viewpoint of aggregation, special properties. Aggregation yields an approximation of accuracy ε^2 in the evaluation of the eigensystem of column block stochastic matrices. For row block stochastic matrices the eigenvalues of the aggregative matrix are also eigenvalues of the original matrix and the eigenvector elements of the aggregative matrix are exact summations of elements of the corresponding eigenvectors of the original matrix. In steady-state analysis this means that a process of aggregation yields exact values for the steady-state probability of being in one aggregate. The remaining error results only from the necessity of approximating the subsystems by matrices that are stochastic. These special properties provide a definition of consistent aggregation in nearly completely decomposable stochastic systems.

CHAPTER III

Criterion for Near-Complete Decomposability Equivalence Classes of Aggregates

There are two independent parts in this chapter. The first copes with the problem of establishing a criterion that would make it possible to determine a priori whether a system enjoys the properties of near-complete decomposability or not. The second clarifies the analogy that exists between the process of aggregation and the process of lumping the states of a Markov chain. These two kinds of results are useful for the applications that are studied in the remaining chapters.

3.1 Necessary Conditions

We assumed in Section 1.1, without more precision, that ε, the maximum degree of coupling between aggregates, is a real positive number, small as compared with the elements of the aggregates $\mathbf{Q}_I{}^*$. So far we have not concerned ourselves with the question of how small ε may have to be; we have not questioned the possibility of the existence of an upper limit for ε beyond which the properties of near-complete decomposability would no longer hold.

This attitude is coherent with the Simon–Ando theorems, which state that whatever standard of approximation is required, there always exists a *sufficiently* small ε so that this standard can be attained. In Chapter II we have made explicit how the degree of approximation is related to ε. This state of affairs might suggest that *all* systems could, in principle, be considered nearly completely decomposable, their maximum degree of coupling and the accuracy with which they can be analyzed by aggregation being their only distinguishable characteristics.

However, the assumptions on which the Simon–Ando theorems are based make it possible to isolate a condition that, if not fulfilled, prevents a system from enjoying the proper short- and long-term equilibrium characteristics of near-complete decomposability.

In Chapter I [Eqs. (1.15) and (1.16)], we specified the dynamic behavior of the original system $\mathbf{Q}$ and of the corresponding aggregate system $\mathbf{Q}^*$ in terms of the idempotent expansions of these matrices as

$$\mathbf{Q}^t = \mathbf{Z}(1_1) + \sum_{I=2}^{N} \lambda^t(1_I)\mathbf{Z}(1_I) + \sum_{I=1}^{N}\sum_{i=2}^{n(I)} \lambda^t(i_I)\mathbf{Z}(i_I),$$

$$\mathbf{Q}^{*t} = \sum_{I=1}^{N} \mathbf{Z}^*(1_I) + \sum_{I=1}^{N}\sum_{i=2}^{n(I)} \lambda^{*t}(i_I)\mathbf{Z}^*(i_I).$$

In Section 1.3, time T_1^* and time T_2 were defined. For $t < T_2$, the values $|\lambda^t(1_I)|$ are supposed to remain close to unity, so that the behavior of both $\mathbf{Q}^t$ and $\mathbf{Q}^{*t}$ is determined by the last terms of their respective expansions, and thereby (Theorem 1.1) is similar. For $t > T_1^*$, the second summation in the expansion of $\mathbf{Q}^*$, i.e., the values $|\lambda^{*t}(i_I)|$, $i \neq 1$, have almost vanished, each aggregate having reached its own internal equilibrium.

Thus, for $T_1^* < t < T_2$, the original system $\mathbf{Q}$ is in a short-term equilibrium close to the equilibrium of the completely decomposable system $\mathbf{Q}^*$. For this to occur, it is clearly necessary that

$$T_1^* < T_2. \tag{3.1}$$

As T_1^* is independent from ε, and T_2 increases beyond limit when $\varepsilon \to 0$, inequality (3.1) can certainly be verified for a sufficiently small ε. However, it is equally clear that, if inequality (3.1) is not verified, there exists no time period during which $\mathbf{Q}$ may be considered to be in an equilibrium approximately identical to that of a completely decomposable system. The verification of (3.1) depends on ε being sufficiently small; how small ε must be is the question we shall try to answer now.

The definitions of T_1^* and T_2 stipulate that, with $T_2 > T_1^*$, both

$$\max_I [1 - |\lambda(1_I)|^{T_2}], \qquad \text{and} \qquad \max_{I, i \neq 1} [|\lambda^*(i_I)|^{T_1^*}] = \max_I [|\lambda^*(2_I)|^{T_1^*}]$$

must be small compared to unity. For this to hold it is *necessary* that

$$\min_I |\lambda(1_I)| > \max_I |\lambda^*(2_I)|. \tag{3.2}$$

This inequality, in fact, specifies the lowest bound for the convergence rate of the slowest aggregate, since each aggregate $\mathbf{Q}_I^*$ converges toward equilibrium at the same speed at which a geometrical progression of quotient $\lambda^*(2_I)$ converges toward zero. Thus, the smaller $\lambda^*(2_I)$, the better the convergence [see, e.g., Bodewig (1956, p. 230 ff)].

3.2 Criterion for Near-Complete Decomposability

As such, the *necessary* condition (3.2) is not of much practical interest since none of its constituents is a priori known. Fortunately, however, for both sides of the inequality bounds can be defined in terms of ε and of the aggregate elements, and a criterion for the near-complete decomposability of a matrix can then be obtained in terms of these bounds. We proceed very much in the same way as in Section 2.5.

It follows from Theorem 2.1 that, if we neglect ε^2, the eigenvalues $\lambda(1_I)$ are also eigenvalues of the $N \times N$ aggregative matrix $\mathbf{P}$. The elements of $\mathbf{P}$ are defined by (1.25) and are such that

$$\min_I (p_{II}) = 1 - \max_I \left(\sum_{J \neq I} p_{IJ} \right) \geqslant 1 - \varepsilon. \tag{3.3}$$

We must again restrict ourselves to systems for which $\varepsilon < \frac{1}{2}$. As already said, this restriction represents no loss of generality whatsoever in stochastic systems; aggregates lose their meaning if the probability of leaving them can attain $\frac{1}{2}$.

If $\varepsilon < \frac{1}{2}$, the aggregation matrix $\mathbf{P}$ is diagonally dominant since, by (3.3), we have

$$\min_I (p_{II}) > \max_I \left(\sum_{J \neq I} p_{IJ} \right).$$

Moreover, each eigenvalue of $\mathbf{P}$ lies in at least one of the Gerschgorin circular disks with center p_{II} and radii $\sum_{J \neq I} |p_{IJ}|$; thus, as $\varepsilon < \frac{1}{2}$, all eigenvalues of $\mathbf{P}$ have a real part that is positive ($\mathbf{P}$ is positive definite), and we have also

$$\min_I |\lambda(1_I)| \geqslant 1 - 2\varepsilon. \tag{3.4}$$

Since $\mathbf{P}$ is also indecomposable, the equality can be verified, according to a theorem by Taussky (1948), for an eigenvalue that would be a common boundary point to all disks, for example, with the matrix

$$\begin{bmatrix} 1-a & a \\ a & 1-a \end{bmatrix}.$$

The eigenvalues of this matrix are 1 and $1-2a$. In terms of ε, $1-2\varepsilon$ is thus the greatest lower bound for $\min_I |\lambda(1_I)|$.

If we introduce (3.4) into (3.2), we conclude that *a sufficient condition for near-complete decomposability is*

$$\varepsilon < \left[1 - \max_I |\lambda^*(2_I)| \right] \Big/ 2, \tag{3.5}$$

this condition being the least coercive one that can be defined in terms of ε. Let us now remember that, in Section 2.2.2, $(1-|\lambda^*(2_I)|)$ was introduced as a lower bound of aggregate $\mathbf{Q}_I^*$ indecomposability. Hence, condition (3.5) relates the maximum degree of coupling between aggregates to the minimum indecomposability of an aggregate. This result is all the more useful due to the fact that by using the upper bounds (2.23) and (2.24) for $|\lambda^*(2_I)|$ that were introduced in Section 2.2.2, one can express condition (3.5) as a function of the aggregate elements.

Condition (3.5) can be easily generalized to multilevel nearly completely decomposable systems. If ε_{l+1} is, as defined in Section 1.5, the maximum degree of coupling between aggregates $\mathbf{P}_I^{*[l]}$, and if $\lambda^{*[l]}(2_I)$ is the second largest eigenvalue of $\mathbf{P}_I^{*[l]}$, *a sufficient condition of near-complete decomposability at level* $l+1$ *is*

$$\varepsilon_{l+1} < \left[1 - \max_I |\lambda^{*[l]}(2_I)|\right] \Big/ 2, \tag{3.6}$$

with I running over the aggregates $\mathbf{P}_I^{*[l]}$.

The upper bound (2.23) or (2.24) of Section 2.2.2 may serve to assess an upper bound for $\max_I |\lambda^{*[l]}(2_I)|$, and thus a *lower bound* of aggregate indecomposability in terms of aggregate elements. If we substitute this lower bound into (3.6), we obtain a sufficient condition of near-complete indecomposability expressed in terms of aggregate elements only.

3.3 Lumping States

We now turn our attention to the second topic treated in this chapter, namely, the problem of defining classes of equivalence for the aggregates. First we must remember what the process of lumping states in a Markov chain consists of.

Assume we are given a n-state homogeneous Markov chain with transition probability matrix $\mathbf{Q}$. Let $A = \{A_1, A_2, \ldots, A_R\}$ be a partition of the set of states of this Markov chain. Each subset A_i, $i = 1, \ldots, R$ can be considered a state of a new process. If we use S_t to denote the state occupied by this new process at time t, the probability of a transition occurring at time t from state A_i to state A_j can be denoted $p_{A_iA_j}(t)$ and is given by

$$p_{A_iA_j}(t) = \Pr\{S_t = A_j \mid S_{t-1} = A_i \text{ and } S_{t-2} = A_k \text{ and } \cdots \text{ and } S_0 = A_m\}. \tag{3.7}$$

The original Markov chain is thus reduced to a stochastic process with fewer states; this procedure can be useful when the analysis does not require the level of detail of the original process. The new process is called a *lumped process* and A_i a *lumped state* (Kemeny and Snell, 1960, pp. 123ff.).

The lumped process is again a first-order homogeneous Markov chain only if $p_{A_iA_j}(t)$ is time independent and depends only on S_{t-1}:

$$p_{A_iA_j}(t) = p_{A_iA_j} = \Pr\{S_t = A_j \mid S_{t-1} = A_i\} \qquad \text{for all} \quad t > 0.$$

Kemeny and Snell qualify a Markov chain as being *lumpable with respect to a partition* $(A_1, \ldots, A_R)$ if for every possible initial state the lumped process defined by (3.7) is a first-order homogeneous Markov chain and does not depend on the choice of the initial state. Defining

$$q_{iA_j} = \sum_{j \in A_j} q_{ij}$$

as the probability of moving from state i to set A_j, Kemeny and Snell prove that *a necessary and sufficient condition for a homogeneous Markov chain to be lumpable with respect to a partition* $(A_1, \ldots, A_R)$ *is that all the* q_{iA_j} *have the same value, say* W_{ij}, *for every* $i \in A_i$, *and for any given* $A_j \neq A_i$.

In other words, to be lumpable, the matrix $\mathbf{Q}$ must be row block stochastic when it is arranged so that the states of a same set A_i have adjacent entries. In this case the transition matrix of the lumped process is the $R \times R$ matrix $[W_{ij}]$. Our Corollary 2.1 is therefore applicable to a lumpable matrix $\mathbf{Q}$. Thus we know that the eigenvalues of the lumped matrix $[W_{ij}]$ are eigenvalues of the original matrix $\mathbf{Q}$, and the ith element of each eigenvector of the lumped matrix is the sum of the elements of the corresponding eigenvector of $\mathbf{Q}$ that correspond to the subset A_i. In particular, the steady-state probability of being in any lumped state A_i of the lumped process is the sum of the steady-state probabilities of being in each individual state that belongs to A_i.

3.4 Classes of Equivalence for Aggregates

3.4.1 Clearly, a lumpable system is a particular case of a nearly completely decomposable system in which the probability of a transition between two aggregates is the same for all departure states. In this case the interactions between aggregates are independent of the interactions within aggregates, and at any instant of its evolution the system may be regarded as a set of aggregates only, *however large the coupling between these aggregates is.*

The interest of the lumping process lies in the fact that we shall later meet systems that are nearly completely decomposable with respect to one partition of the states and at the same time lumpable with respect to another. This is especially interesting when, in two or more aggregates of the same size, all corresponding states can be lumped together; the set of lumped states then becomes the set of states of a new aggregate. This new aggregate presents itself to the other parts of the system with the same behavior as the original aggregates, so that at the upper levels of aggregation where only interactions between aggregates are considered, only this new aggregate needs to be taken into account. All the original aggregates will be said to belong to the same class of equivalence defined by the new aggregate.

This equivalence relation can be more precisely introduced as follows. We assume, without loss of generality, that the matrix $\mathbf{Q}$ is nearly completely decomposable with respect to the subsystems $\mathbf{Q}_1, \mathbf{Q}_2, \ldots, \mathbf{Q}_L, \ldots, \mathbf{Q}_N$, where $\mathbf{Q}_1, \ldots, \mathbf{Q}_L$ have the same size $n(1) = n(2) = \cdots = n(L) = r$. Now, let (A_i, k_K), $i = 1, \ldots, r$, $k = 1, \ldots, n(K)$, $K = L+1, \ldots, N$, be a partition of the states of $\mathbf{Q}$ with

$$A_i = \{i_1, i_2, \ldots, i_L\}, \qquad i = 1, \ldots, r.$$

This partition consists therefore of all individual states k_K of subsystems $\mathbf{Q}_{L+1}, \ldots, \mathbf{Q}_N$, and of the r sets of the state i, $i = 1, \ldots, r$, of each subsystem $\mathbf{Q}_1, \ldots, \mathbf{Q}_L$. States that belong to the same subset of the partition will sometimes be called *corresponding states.*

We shall say that the subsystems $\mathbf{Q}_1, \ldots, \mathbf{Q}_L$ belong to the same class of equivalence if the matrix $\mathbf{Q}$ is lumpable with respect to the partition (A_i, k_K). This class of equivalence is itself a subsystem with states $A_1, \ldots, A_r$; the $r \times r$ transition matrix of this class of equivalence has off-diagonal elements equal (for some $I = 1, \ldots, L$) to

$$q_{A_i A_j} = \sum_{J=1}^{L} q_{i_I j_J}, \qquad i \neq j, \tag{3.8}$$

and the transition probability from a state A_i of the equivalence class to a state k_K of another subsystem $\mathbf{Q}_K$ is given (for some $I = 1, \ldots, L$) by

$$q_{A_i k_K} = q_{i_I k_K}, \qquad k = 1, \ldots, n(K), \quad K = L+1, \ldots, N. \tag{3.9}$$

Thus, the following theorem is a direct result of the Kemeny–Snell lumpability conditions:

Theorem 3.1 *The necessary and sufficient conditions for* $\mathbf{Q}$ *to be lumpable with respect to the partition* (A_i, k_K)*, and thus for the subsystems* $\mathbf{Q}_1, \ldots, \mathbf{Q}_L$ *to belong to the same class of equivalence is that the probabilities* $\sum_{J=1}^{L} q_{i_I j_J}$, $i \neq j$, *have the same value* $q_{A_i A_j}$ *for* $I = 1, \ldots, L$, *and that the probabilities* $q_{i_I k_K}$, $K = L+1, \ldots, N$, *have the same value* $q_{A_i k_K}$ *for* $I = 1, \ldots, L$.

The lumpability condition is necessarily verified for all the other probabilities

$$q_{k_K A_i} = \sum_{I=1}^{L} q_{k_K i_I}, \tag{3.10}$$

since each state k_K alone is a subset of the partition. In other words, Theorem 3.1 specifies conditions for the output behavior of the subsystems with respect to the remainder of the system, but imposes no condition whatsoever on their input from the remainder of the system.

So far, we have defined classes of equivalence in terms of subsystems only; but clearly if one or more subsystems belong to a same equivalence class, so do the aggregates that can be derived from these subsystems. So from now on we shall also speak freely of *equivalence classes of aggregates*.

3.4.2 The advantage of grouping aggregates into equivalence classes is twofold.

First, and rather obviously, the number of aggregates that have to be taken into consideration at the upper levels of aggregation is reduced. If the matrix $\mathbf{Q}$ is nearly completely decomposable with respect to the subsystems $\mathbf{Q}_1, \ldots, \mathbf{Q}_L, \ldots, \mathbf{Q}_N$, the equivalence class $\mathbf{Q}_1, \ldots, \mathbf{Q}_L$ introduced above can be

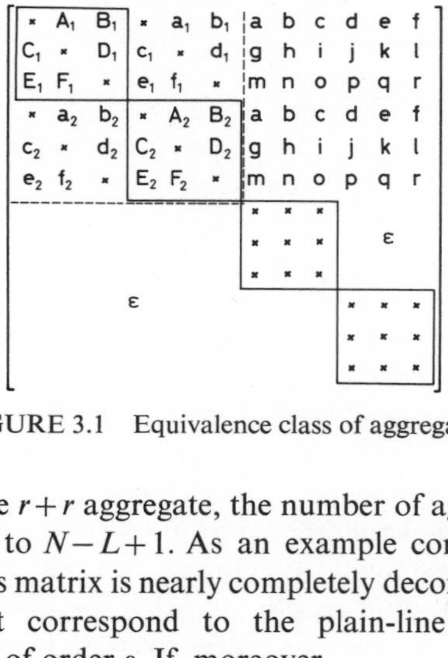

$$\begin{bmatrix} * & A_1 & B_1 & * & a_1 & b_1 & a & b & c & d & e & f \\ C_1 & * & D_1 & c_1 & * & d_1 & g & h & i & j & k & l \\ E_1 & F_1 & * & e_1 & f_1 & * & m & n & o & p & q & r \\ * & a_2 & b_2 & * & A_2 & B_2 & a & b & c & d & e & f \\ c_2 & * & d_2 & C_2 & * & D_2 & g & h & i & j & k & l \\ e_2 & f_2 & * & E_2 & F_2 & * & m & n & o & p & q & r \\ & & & & & & * & * & * & & & \\ & & & & & & * & * & * & & \varepsilon & \\ & & & & & & * & * & * & & & \\ & & & & \varepsilon & & & & & * & * & * \\ & & & & & & & & & * & * & * \\ & & & & & & & & & * & * & * \end{bmatrix}$$

FIGURE 3.1 Equivalence class of aggregates.

represented by a single $r+r$ aggregate, the number of aggregates being in this case reduced from N to $N-L+1$. As an example consider the matrix displayed in Fig. 3.1. This matrix is nearly completely decomposable into the four 3×3 aggregates that correspond to the plain-line partition if all off-diagonal elements are of order ε. If, moreover,

$$A_1 + a_1 = A_2 + a_2, \qquad B_1 + b_1 = B_2 + b_2, \quad \ldots, \quad E_1 + e_1 = E_2 + e_2,$$

then the first two subsystems on the main diagonal belong to the same class of equivalence. Their corresponding states can be lumped together so that only three 3×3 aggregates will subsist in the lumped matrix. Correspondingly, the aggregative matrix of interaggregate transitions that must be considered is only a 3×3 matrix, and the first element of the steady-state vector of this aggregative matrix is the steady-state probability of occupying any state of the equivalence class. This probability must then be multiplied by every element of the steady-state vector of each aggregate that belongs to the equivalence class. Every product so obtained is an approximation to the steady-state probability of being in a particular state of the equivalence class.

The second advantage of grouping aggregates into equivalence classes is that the maximum degree of coupling that must be taken into account is also reduced; indeed, the coupling between subsystems that belong to the same equivalence class can be neglected when this equivalence class is considered as one aggregate only. More precisely, the probability $q_{A_iA_j}$ of a transition between two states of the *same* equivalence class that is defined by (3.8) can be rewritten

$$q_{A_iA_j} = q_{i_Ij_I} + \sum_{J=1, J\neq I}^{L} q_{i_Ij_I}, \qquad i \neq j.$$

The summation on the right-hand side is the probability of a transition between *different* subsystems $\mathbf{Q}_I$, $I = 1, \ldots, L$, and thereby would be part of the maximum degree of coupling between aggregates if each individual subsystem $\mathbf{Q}_I$ were considered a separate aggregate.

3.4.3 ***Nearly Lumpable Matrices*** So far, we have considered only the case of lumpable matrices that reduce to lumped matrices that are nearly completely decomposable. We shall also encounter *nearly lumpable* matrices, i.e., matrices $\mathbf{Q}$ that can be written in the form

$$\mathbf{Q} = \mathbf{Q}' + \varepsilon\mathbf{C}',$$

where $\mathbf{Q}'$ is a lumpable matrix. If the lumped matrix obtained from $\mathbf{Q}'$ is completely decomposable into equivalence classes, then ε is the maximum degree of coupling between these classes, and $\mathbf{Q}$ is nearly completely decomposable into these classes. This would be, for example, the case of the matrix in Fig. 3.1 if the sums $A_1 + a_1$ and $A_2 + a_2$, $B_1 + b_1$ and $B_2 + b_2$, etc., were not rigorously equal but differed by at most ε.

In the two cases (i.e., lumpable matrices that reduce to nearly completely decomposable matrices, and nearly lumpable matrices that reduce to completely decomposable matrices) the process of analysis by aggregation remains the same.

CHAPTER IV

Decomposability of Queueing Networks

This chapter starts the second part of this monograph in which the concept of near-complete decomposability is exploited to analyze stochastic networks of interconnected queues. The motivations of this analysis were given in the introduction. We shall first investigate, as generally as possible, the conditions under which queueing networks enjoy the property of near-complete decomposability. Then, in Chapters V and VI, we shall see how networks that satisfy these conditions may be approximated by a technique of variable aggregation.

We have demonstrated that the approximation made by aggregation is essentially of the same order of magnitude as the maximum degree of coupling ε between aggregates. So our first goal will be to determine what the aggregate structure of queueing networks is, and to express the maximum degree of coupling between aggregates in function of the parameters of the network. We shall then be able to determine a priori the degree of approximation that is to be expected for a specific network analyzed by aggregation.

To make the presentation as clear as possible, without, however, neglecting any interesting generality, we first study the conditions under which a simple model of a network is nearly completely decomposable, and postpone to Chapter VI the study of these conditions for a more general class of networks.

4.1 Basic Model

Our basic model is a typical case of the class of queueing networks studied by Jackson (1963) and Gordon and Newell (1967). We consider a set of $L+1$ resources $R_0, R_1, \ldots, R_L$, each providing a certain type of service (Fig. 4.1). The service time of resource R_l is supposed to be a random variable geometrically distributed with parameter μ_l; the probability that this service requires τ discrete time units is thus $(1 = \mu_l)^{\tau-1}\mu_l$, and the mean service time is $1/\mu_l$. We shall refer to μ_l as the *service rate* of resource R_l.

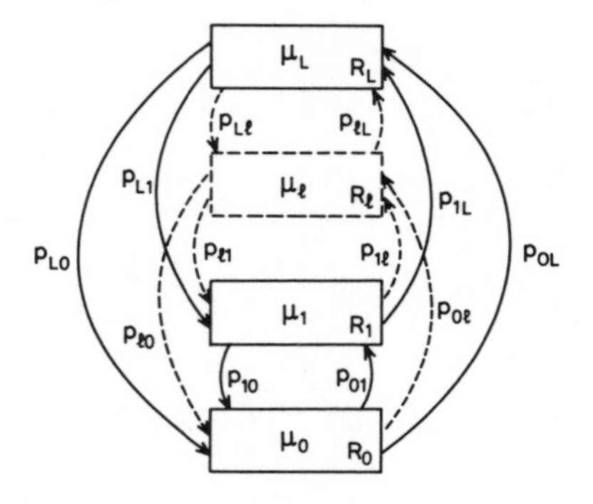

FIGURE 4.1 Basic model of a queueing network.

The behavior of a customer in this system corresponds to a random excursion among the resources; a customer who completes service at resource R_m applies immediately to resource R_l (which may be R_m) with probability p_{ml}, $0 \leqslant m, l \leqslant L$. For $m = 0, \ldots, L$, $\sum_{l=0}^{L} p_{ml} = 1$.

We designate by N, $N < \infty$, the total number of customers in the system, the behavior of each customer being specified by the same set of transfer probabilities p_{ml}. If we let i_l, $l = 0, \ldots, L$, be the number of customers in service or in queue at resource R_l, we have

$$\sum_{l=0}^{L} i_l = N, \tag{4.1}$$

since customers may at most be waiting for or being served by one resource at a time. The value N is supposed to remain constant in the system. This is not so important a restriction as might appear at first sight: if necessary, an equivalent system may be conceived in which the total number of customers $\sum_{l=0}^{L} i_l$ would vary in time without exceeding N, by simply considering an additional resource R_{L+1} of service rate

$$\mu_{L+1} = \lambda \times i_{L+1}, \qquad i_{L+1} = N - \sum_{l=0}^{L} i_l, \tag{4.2}$$

which would model the customer arrivals. In this case $p_{(L+1)l}$ and $p_{l(L+1)}$ would be the probabilities of resource R_l, $l = 0, \ldots, L$, being the first and the last, respectively, to be requested by a customer.

The state of the system is uniquely defined by the $(L+1)$-tuple $(i_0, \ldots, i_L)$, and there are $\binom{L+N}{L}$ such distinguishable states, i.e., the number of partitions of N customers among $L+1$ sets.

Let $p(i_0, \ldots, i_L, t)$ denote the joint probability that at time t the system is in state $(i_0, \ldots, i_L)$. If we choose the time unit small enough so that the probability of more than one service being completed during this time unit may be neglected, these probabilities satisfy the system of linear equations (Jackson,

1963; Gordon and Newell, 1967)

$$\begin{aligned} p(i_0, i_1, \ldots, i_L, t+1) &= p(i_0, i_1, \ldots, i_L, t)\left[1 - \sum_{l=0}^{L} \kappa(i_l)\,\mu_l(1-p_{ll})\right] \\ &\quad + \sum_{l=0}^{L} \sum_{m=0, m\neq l}^{L} \kappa(i_m)\,p(i_0, \ldots, i_l+1, \ldots, i_m-1, \ldots, i_L, t)\,\mu_l p_{lm}, \end{aligned} \tag{4.3}$$

where the binary function

$$\kappa(i_l) = \begin{cases} 0 & \text{if} \quad i_l = 0, \\ 1 & \text{if} \quad i_l \neq 0, \end{cases}$$

accounts for the impossibility of any i_l taking negative values.

These equations state that the only possible transitions during a time unit are for all l, m, $l \neq m$, from state $(i_0, \ldots, i_l+1, \ldots, i_m-1, \ldots, i_L)$ to state $(i_0, \ldots, i_l, \ldots, i_m, \ldots, i_L)$ with probability $\mu_l p_{lm}$, and from this latter state into itself with probability

$$1 - \sum_{l=0}^{L} \sum_{m=0, m\neq l}^{L} \kappa(i_l)\,\mu_l p_{lm} = 1 - \sum_{l=0}^{L} \kappa(i_l)\,\mu_l(1-p_{ll}). \tag{4.4}$$

Written in matrix form, Eq. (4.3) becomes

$$\mathbf{p}(t+1) = \mathbf{p}(t)\mathbf{Q}, \tag{4.5}$$

where $\mathbf{p}(t)$ denotes the probability vector of elements $p(i_0, \ldots, i_L, t)$, and $\mathbf{Q}$ is a Markov transition probability matrix of size $\binom{L+N}{L}$. All nondiagonal elements of $\mathbf{Q}$ are either zero or equal to some probability $\mu_l p_{im}$, and all diagonal elements have the form (4.4).

The equilibrium probability distribution of the system

$$\mathbf{p} = \lim_{t\to\infty} \mathbf{p}(t) \tag{4.6}$$

is by definition the solution of the system $\mathbf{p} = \mathbf{pQ}$.

We shall assume that the Markov chain so defined is irreducible, viz., that each state can be reached from any other state by a finite series of transitions. In this case, $\mathbf{Q}$ is a nonnegative indecomposable matrix. Thus, we know by the Perron–Frobenius theorem [see, e.g., Marcus and Minc (1964, p. 124)] that the equilibrium probability distribution $\mathbf{p}$ exists and is the positive eigenvector corresponding to the maximal eigenvalue of $\mathbf{Q}$, which is equal to unity.

4.2 Conditions for Near-Complete Decomposability

Each index value j, $j = 1, \ldots, \binom{L+N}{L}$, of the square stochastic matrix $\mathbf{Q}$ refers to a distinct state $(i_0, i_1, \ldots, i_L)$ of the system. We choose to number these states according to the lexicographic ordering; more precisely, we give increasing index values to states that yield increasing values for the sum $\sum_{l=0}^{L} i_l N^l$.

$i_0i_1i_2i_3$	2000	1100	0200	1010	0110	0020	1001	0101	0011	0002
2000	$1-\Sigma_1$	$\mu_0 p_{01}$	0	$\mu_0 p_{02}$	0	0	$\mu_0 p_{03}$	0	0	0
1100	$\mu_1 p_{10}$	$1-\Sigma_2$	$\mu_0 p_{01}$	$\mu_1 p_{12}$	$\mu_0 p_{02}$	0	$\mu_1 p_{13}$	$\mu_0 p_{03}$	0	0
0200	0	$\mu_1 p_{10}$	$1-\Sigma_3$	0	$\mu_1 p_{12}$	0	0	$\mu_1 p_{13}$	0	0
1010	$\mu_2 p_{20}$	$\mu_2 p_{21}$	0	$1-\Sigma_4$	$\mu_0 p_{01}$	$\mu_0 p_{02}$	$\mu_2 p_{23}$	0	$\mu_0 p_{03}$	0
0110	0	$\mu_2 p_{20}$	$\mu_2 p_{21}$	$\mu_1 p_{10}$	$1-\Sigma_5$	$\mu_1 p_{12}$	0	$\mu_2 p_{23}$	$\mu_1 p_{13}$	0
0020	0	0	0	$\mu_2 p_{20}$	$\mu_2 p_{21}$	$1-\Sigma_6$	0	0	$\mu_2 p_{23}$	0
1001	$\mu_3 p_{30}$	$\mu_3 p_{31}$	0	$\mu_3 p_{32}$	0	0	$1-\Sigma_7$	$\mu_0 p_{01}$	$\mu_0 p_{02}$	$\mu_0 p_{03}$
0101	0	$\mu_3 p_{30}$	$\mu_3 p_{31}$	0	$\mu_3 p_{32}$	0	$\mu_1 p_{10}$	$1-\Sigma_8$	$\mu_1 p_{12}$	$\mu_1 p_{13}$
0011	0	0	0	$\mu_3 p_{30}$	$\mu_3 p_{31}$	$\mu_3 p_{32}$	$\mu_2 p_{20}$	$\mu_2 p_{21}$	$1-\Sigma_9$	$\mu_2 p_{23}$
0002	0	0	0	0	0	0	$\mu_3 p_{30}$	$\mu_3 p_{31}$	$\mu_3 p_{32}$	$1-\Sigma_{10}$

FIGURE 4.2 Matrix $\mathbf{Q}(N, L)$, $N = 2$, $L = 3$.

An example of a matrix with rows and columns arranged in this order is given in Fig. 4.2.

For $N \geqslant 0$ and $L \geqslant 1$, we shall for convenience sake denote $\mathbf{Q}(N, L)$ the lexicographically arranged matrix of system (4.5). If we remember that its diagonal elements $q_{jj}(N, L)$ take the form (4.4), and that all off-diagonal elements $q_{jk}(N, L)$, $j \neq k$, are either 0 or equal to some probability $\mu_l p_{lm}$, $l \neq m$, we can prove the following:

Lemma 4.1 *The stochastic matrix $\mathbf{Q}(N, L)$, $N > 0$, $L > 1$, may be partitioned among $N+1$ principal submatrices $\mathbf{Q}'(i_L)$, $i_L = 0, \ldots, N$. Each submatrix $\mathbf{Q}'(i_L)$ is, except for its main diagonal, identical to a matrix $\mathbf{Q}(N-i_L, L-1)$.*

PROOF Because of the lexicographic ordering of the system states, the smaller index values of $\mathbf{Q}(N, L)$ refer to *all* states with $i_L = 0$; there are $\binom{L-1+N}{L-1}$ such states, i.e., the number of distinct integer solutions of equation $\sum_{l=0}^{L-1} i_l = N$. Likewise, the $\binom{L-1+N-1}{L-1}$ following index values refer to all states with $i_L = 1$, and there are as many such states as there are solutions of the equation $\sum_{l=0}^{L-1} i_l = N-1$; and so on, up to $i_L = N$. All states may therefore be partitioned among $N+1$ sets, $i_L = 0, \ldots, N$, the (i_L+1)th set grouping all the $\binom{L-1+N-i_L}{L-1}$ states for which

$$\sum_{l=0}^{L-1} i_l = N - i_L,$$

in agreement with the identity [see, e.g., Riordan (1968, p. 7)]:

$$\sum_{i_L=0}^{N} \binom{L-1+N-i_L}{L-1} = \binom{L+N}{L}.$$

These $N+1$ sets define as many principal submatrices all of whose nondiagonal elements $\mu_l p_{lm}$ are the transition probabilities between any two distinct states of the set of all states for which i_L has a given value. This set is the complete set of states of a system $\mathbf{Q}(N-i_L, L-1)$ consisting of resources $R_0, \ldots, R_{L-1}$, with a finite population of $N-i_L$ customers.

In order to help visualize the structure of matrix $\mathbf{Q}(N, L)$, matrix $\mathbf{Q}(2,3)$ is displayed in Fig. 4.2, where $\sum_i$ denotes the ith row sum of nondiagonal elements. Solid lines isolate submatrices $\mathbf{Q}'(i_3)$, $i_3 = 0, 1, 2$.

Now, if we recursively apply Lemma 4.1 to the submatrices $\mathbf{Q}'(i_L)$, we obtain the central theorem of this chapter:

Theorem 4.1 *If, for* $l = 1, \ldots, L-1$,

$$\omega_l = \max_{\substack{i_0, \ldots, i_L \\ \sum_l i_l = N}} \left[\sum_{k=l+1}^{L} \kappa(i_k)\, \mu_k \sum_{m=0}^{l} p_{km} + \sum_{k=0}^{l} \kappa(i_k)\, \mu_k \sum_{m=l+1}^{L} p_{km} \right] \tag{4.7}$$

is sufficiently small, then the stochastic matrix $\mathbf{Q}(N, L)$, $N > 0$, $L > 1$, *defines a* $(L-1)$*-level nearly completely decomposable system. Each level of aggregation* l *consists of* $\binom{L-l+N}{L-l}$ *aggregates, which belong to* $N+1$ *classes of equivalence* $\mathbf{Q}(n, l)$, $n = 0, \ldots, N$. *The maximum degree of coupling between these classes is* ε_l, *and is at most equal to* ω_l.

PROOF Since a matrix $\mathbf{Q}'(i_L)$ is, except for its main diagonal, identical to a matrix $\mathbf{Q}(N-i_L, L-1)$, we can apply Lemma 4.1 to it. If we do this recurrently $L-l$ times, we partition $\mathbf{Q}(N, L)$ into a set of principal submatrices $\mathbf{Q}'(i_{l+1}, \ldots, i_L)$, each such submatrix grouping all states for which $i_{l+1}, \ldots, i_L$ have a given value. There will be

$$\sum_{n=0}^{N} \binom{L-(l+1)+n}{L-(l+1)} = \binom{L-l+N}{L-l}$$

such submatrices $\mathbf{Q}'(i_{l+1}, \ldots, i_L)$, viz., as many as the number of integer solutions of the $N+1$ equations $\sum_{k=l+1}^{L} i_k = N-n$, $n = 0, \ldots, N$; and any of these submatrices is identical, except for the diagonal, to one of the $N+1$ matrices $\mathbf{Q}(n, l)$, with $0 \leqslant n = N - \sum_{k=l+1}^{L} i_k \leqslant N$. These principal submatrices $\mathbf{Q}'(i_{l+1}, \ldots, i_L)$ are lexicographically ordered with respect to $(i_{l+1}, \ldots, i_L)$ along the diagonal; their diagonal elements are those of $\mathbf{Q}(N, L)$, and are for each state yielded by (4.4):

$$1 - \sum_{k=0}^{L} \sum_{m=0, m \neq k}^{L} \kappa(i_k)\, \mu_k\, p_{km},$$

which may be decomposed into $1-(S_1+S_2+S_3+S_4)$ with

$$\begin{aligned} S_1 &= \sum_{k=0}^{l} \sum_{m=0, m \neq k}^{l} \kappa(i_k)\, \mu_k\, p_{km}, & S_2 &= \sum_{k=l+1}^{L} \sum_{m=l+1, m \neq k}^{L} \kappa(i_k)\, \mu_k\, p_{km}, \\ S_3 &= \sum_{k=0}^{l} \sum_{m=l+1}^{L} \kappa(i_k)\, \mu_k\, p_{km}, & S_4 &= \sum_{k=l+1}^{L} \sum_{m=0}^{l} \kappa(i_k)\, \mu_k\, p_{km}. \end{aligned} \tag{4.8}$$

For each possible set of values $(i_{l+1}, \ldots, i_L)$, $\sum_{k=l+1}^{L} i_k = N-n$, i.e., for each submatrix $\mathbf{Q}'(i_{l+1}, \ldots, i_L)$, expression $(1-S_1)$ yields the diagonal elements of a matrix $\mathbf{Q}(n, l)$. Let us then define $\mathbf{C}(N, L-l)$ as a matrix of the same size as $\mathbf{Q}(N, L)$, with diagonal elements yielded by $-(S_3+S_4)$ and off-diagonal elements that are zero except, on each row, those of $\mathbf{Q}(N, L)$, which are the terms of (S_3+S_4). Setting

$$\mathbf{Q}(N, L) = \mathbf{D}(N, L-l) + \mathbf{C}(N, L-l), \tag{4.9}$$

we now obtain a matrix $\mathbf{D}(N, L-l)$ that has $\binom{L-l+N}{L-l}$ principal submatrices $\mathbf{D}'(i_{l+1}, \ldots, i_L)$, each of these submatrices being identical, except for the main diagonal, to one of the $N+1$ matrices $\mathbf{Q}(n, l)$. Outside these submatrices, all elements of $\mathbf{D}(N, L-l)$ are zero except those which, on each row, are the terms of S_2. But S_2 is the probability of a transition from a given state

$$(i_0, \ldots, i_l, \ldots, i_k, \ldots, i_m, \ldots, i_L)$$

to *any* of the states

$$(i_0, \ldots, i_l, \ldots, i_k - 1, \ldots, i_m + 1, \ldots, i_L), \qquad l+1 \leqslant k, \qquad m \leqslant L,$$

all these states having the *same* pattern $(i_0, \ldots, i_l)$ and *same* sum $\sum_{k=l+1}^{L} i_k = N-n$. Therefore, S_2 is the probability of a transition between *corresponding* states of subsystems almost identical to the same matrix $\mathbf{Q}(n, l)$; all transition probabilities between *noncorresponding* states of these subsystems of $\mathbf{D}(N, L-l)$ are therefore zero. All the subsystems that are identical, except for the main diagonal, to the same matrix $\mathbf{Q}(n, l)$ obey therefore the Kemeny–Snell conditions of Theorem 3.1 to lump together their corresponding states; these subsystems belong to the same equivalence class $\mathbf{Q}(n, l)$. The maximum probability ω_l of leaving a subset of elementary states that corresponds to such an equivalence class is equal to the maximum row sum of the off-diagonal elements of $\mathbf{C}(N, L-l)$, i.e., to $\max(S_3+S_4)$; ω_l is an upper bound for ε_l. This completes the proof.

In Fig. 4.2, plain lines isolate submatrices $\mathbf{Q}(n, 2)$, $n = 2, 1, 0$, and dashed lines submatrices $\mathbf{Q}(n, 1)$, $n = 2, 1, 0, 1, 0$.

From the above theorem follows immediately:

Corollary 4.1 *If $N \geqslant L+1$, the maximum degree of coupling at each level of aggregation l is at most equal to*

$$\omega_l = \sum_{k=l+1}^{L} \mu_k \sum_{m=0}^{l} p_{km} + \sum_{k=0}^{l} \mu_k \sum_{m=l+1}^{L} p_{km}. \tag{4.10}$$

The property that, at level l, $\mathbf{Q}(N, L)$ decomposes into $\binom{N+L-l}{L-l}$ aggregates grouped into only $N+1$ distinct classes of equivalence is instrumental in the analysis of a network such as $\mathbf{Q}(N, L)$. It explains why this analysis, as we shall observe it in the following chapter, requires at each level l the knowledge of the distribution of customers only among resources $R_0, \ldots, R_l$ and for populations of fixed size n, $0 \leqslant n \leqslant N$, regardless of the distribution of the $N-n$ remaining customers among the $L-l+1$ remaining resources.

4.3 Sufficient Conditions of Network Decomposability

So far, we have shown how our model of a queueing network is nearly completely decomposable into subnetworks of the same structure, assuming without more precision that the maximum degree of coupling between aggregates was sufficiently small. It is now possible to use the criterion defined in Chapter III to obtain a sufficient condition for the near-complete decomposability of our model in terms of the elements $\mu_l p_{lm}$ of the matrix $\mathbf{Q}(N, L)$. For a system to be nearly completely decomposable this criterion requires the maximum degree of coupling to be less than half the lower bound of aggregate indecomposability; this lower bound is equal to $1 - \max_I |\lambda^*(2_I)|$, where $\lambda^*(2_I)$ is the second largest eigenvalue of aggregate $\mathbf{Q}_I^*$.

Let us first consider the queueing network $\mathbf{Q}(N, 1)$ that consists of two resources R_0 and R_1 only. The transition matrix $\mathbf{Q}(N, 1)$ is of order $N+1$ and its entries correspond to the system states (i_0, i_1). The only possible transitions during a single time unit are from state (i_0, i_1) to state $(i_0 - 1, i_1)$ or to state state $(i_0 + 1, i_1)$. Therefore, since its entries are lexicographically ordered, the matrix $\mathbf{Q}(N, 1)$ is necessarily of Jacobi tridiagonal form

$$\mathbf{Q}(N,1) = \begin{bmatrix} 1-\Sigma_1 & \mu_0 p_{01} & 0 & & & & \\ \mu_1 p_{10} & 1-\Sigma_2 & \mu_0 p_{01} & 0 & & & \\ 0 & \mu_1 p_{10} & 1-\Sigma_3 & \mu_0 p_{01} & 0 & & \\ \cdots & \cdots & \cdots & \cdots & \cdots & & \\ & 0 & \mu_1 p_{10} & 1-\Sigma_{N-1} & \mu_0 p_{01} & 0 & \\ & & 0 & \mu_1 p_{10} & 1-\Sigma_N & \mu_0 p_{01} & \\ & & & 0 & \mu_1 p_{10} & 1-\Sigma_{N+1} & \end{bmatrix},$$

where Σ_i stands for the corresponding row sum of the nondiagonal elements. Moreover, since $\mu_1 p_{10} \times \mu_0 p_{01} > 0$, this matrix is *quasi-symmetric* (Wilkinson, 1965, p. 335), and all its eigenvalues are real and simple (Marcus and Minc, 1964, p. 106). In Appendix IV, it is shown that, for $\mathbf{Q}(N, 1)$, the eigenvalues different from unity are yielded by

$$1 - (\mu_1 p_{10} + \mu_0 p_{01}) + 2(\mu_1 p_{10} \times \mu_0 p_{01})^{1/2} \cos[K\pi/(N+1)], \qquad K = 1, \ldots, N. \quad (4.11)$$

We therefore obtain the following lemma:

Lemma 4.2 *The indecomposability of the matrix* $\mathbf{Q}(N, 1)$ *is equal to*

$$\mu_1 p_{10} + \mu_0 p_{01} - 2(\mu_1 p_{10} \times \mu_0 p_{01})^{1/2} \cos[\pi/(N+1)]. \tag{4.12}$$

PROOF The second largest eigenvalue of $\mathbf{Q}(N, 1)$ in module is obtained for $K = 1$, and the indecomposability of $\mathbf{Q}(N, 1)$ is equal to one minus this eigenvalue.

Let us now evaluate the indecomposability of the matrix $\mathbf{Q}(n, l)$, $n = 0, \ldots, N$ of an aggregate at any level l, $l = 1, \ldots, L$. As may be observed on Fig. 4.2, this matrix is not tridiagonal for $l > 1$; we must therefore evaluate its indecomposability by a different approach than for $\mathbf{Q}(N, 1)$.

A matrix $\mathbf{Q}(n, l)$ is (Theorem 4.1) nearly completely decomposable into $n+1$ subaggregates $\mathbf{Q}(m, l-1)$, $m = 0, \ldots, n$.

During a single time unit, transitions are possible only between states of two adjacent subaggregates $\mathbf{Q}(m, l-1)$ and $\mathbf{Q}(m-1, l-1)$, $m = n, \ldots, 1$: the transition from a state $(i_0, \ldots, i_{l-1})$ of $\mathbf{Q}(m, l-1)$ to any state of $\mathbf{Q}(m-1, l-1)$ has a probability equal to

$$\sum_{k=0}^{l-1} \kappa(i_k)\, \mu_k\, p_{kl}, \tag{4.13}$$

while the probability of a transition from *any* state of $\mathbf{Q}(m-1, l-1)$ to any state of $\mathbf{Q}(m, l-1)$ is given by

$$B_l = \mu_l \sum_{k=0}^{l-1} p_{lk}. \tag{4.14}$$

Probability B_l is independent of the state of departure in $\mathbf{Q}(m-1, l-1)$, while probability (4.13) is a function of the state of departure in $\mathbf{Q}(m, l-1)$. Let us then consider a stochastic matrix identical to $\mathbf{Q}(n, l)$ except that probability (4.13) would be equal to

$$A_l = \min_{1 \leqslant \sum_k i_k = m \leqslant n} \left[\sum_{k=0}^{l-1} \kappa(i_k)\, \mu_k\, p_{kl} \right] = \min_{0 \leqslant k \leqslant l-1} (\mu_k p_{kl}). \tag{4.15}$$

This matrix would be row block stochastic with all block row sums equal to A_l above the diagonal and to B_l below. We can assume that the indecomposability of this block-stochastic matrix, in which, *ceteris paribus*, only some elements above the string of diagonal blocks have been made smaller and the corresponding diagonal elements have been made larger, is less or at most equal to the indecomposability of $\mathbf{Q}(n, l)$. Moreover, since it is block stochastic, the $n+1$ largest eigenvalues of this block-stochastic matrix are (Corollary 2.1) the same as those of its associated *aggregative matrix* of transitions between the subaggregates $\mathbf{Q}(m, l-1)$; thus, these two matrices have also the same indecomposability. This aggregative matrix is tridiagonal with all elements above the diagonal equal to A_l, and below equal to B_l; its indecomposability is thus (Lemma 4.2) equal to

$$A_l + B_l - 2(A_l B_l)^{1/2} \cos[\pi/(n+1)], \tag{4.16}$$

and this is also a lower bound for the indecomposability of $\mathbf{Q}(n,l)$. This lower bound is clearly minimum for $n = N$. We therefore conclude that *at each level of aggregation l, $l = 1, 2, \ldots, L-1$, $\mathbf{Q}(N,l)$ is the equivalence class of aggregates with minimum indecomposability, and this minimum indecomposability is at least equal to*

$$A_l + B_l - 2(A_l B_l)^{1/2} \cos[\pi/(N+1)] \tag{4.17}$$

with

$$A_l = \min_{0 \leqslant k \leqslant l-1} (\mu_k p_{kl}), \tag{4.18}$$

$$B_l = \mu_l \sum_{k=0}^{l-1} p_{kl}. \tag{4.19}$$

We are able to determine only a lower bound for the indecomposability of matrices $\mathbf{Q}(n,l)$, $l > 1$; the reason is that formal expressions of the eigenvalues of a matrix in terms of elements of this matrix are hard to obtain for matrices other than tridiagonal that have almost all elements equal along each diagonal. When this is verified, as in $\mathbf{Q}(N,1)$, the characteristic equation can be solved (see Appendix IV) by means of a system of recurrent difference equations with constant coefficients. Our lower bound was obtained on the assumption that decreasing certain off-diagonal elements outside the aggregates in a nearly completely decomposable matrix $\mathbf{Q}(n,l)$ may only decrease the maximum degree of coupling between these aggregates, and thus decrease the matrix indecomposability; remember indeed that $\max_{I \neq 1} \lambda(1_I) \to 1$ as $\varepsilon \to 0$.

Having a lower bound of the indecomposability of each equivalence class of aggregates, we can now make use of the criterion for near-complete decomposability defined in Chapter III in order to arrive at sufficient conditions for the near-complete decomposability of $\mathbf{Q}(N,L)$:

For the stochastic matrix $\mathbf{Q}(N,L)$, $N > 0$, $L > 1$, to be $(L-1)$-level nearly completely decomposable, with each level of aggregation l consisting of $N+1$ distinct equivalence classes of aggregates $\mathbf{Q}(n,l)$, $n = N, \ldots, 0$, it is sufficient that for $l = 1, \ldots, L-1$

$$\omega_l \leqslant \tfrac{1}{2}(A_l + B_l) - (A_l B_l)^{1/2} \cos[\pi/(N+1)], \tag{4.20}$$

where ω_l, A_l, and B_l are defined by (4.7) [or (4.10)], (4.18), *and* (4.19), *respectively.*

These conditions follow immediately from Theorem 4.1, expression (4.17), and the criterion of Chapter III.

4.4 Discussion

These sufficient conditions for the near-complete decomposability of $\mathbf{Q}(N, L)$ are in fact rather severe. This is best illustrated by the following simple case. Consider in Fig. 4.3 the network $\mathbf{Q}(N, 2)$ consisting of resources R_0, R_1, and R_2.

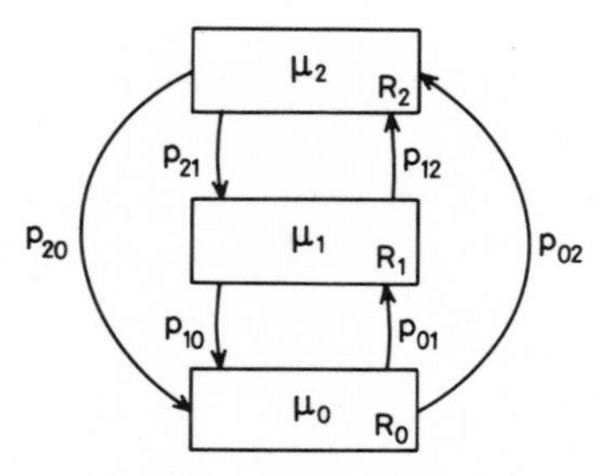

FIGURE 4.3 Network $\mathbf{Q}(N, 2)$.

At the unique level of aggregation ($l = 1$), the aggregates are the systems $\mathbf{Q}(n, 1), n = N, \ldots, 0$, which consist of resources R_0 and R_1, and their maximum degree of coupling ε_1 is the probability of a transition between (R_0, R_1) on the one hand and R_2 on the other. Thus, if we assume without significant loss of generality that $N \geqslant L+1$, we have

$$\varepsilon_1 = \omega_1 = \mu_2(p_{21}+p_{20}) + \mu_0 p_{02} + \mu_1 p_{12}.$$

If we further suppose that $\mu_0 p_{01} \simeq \mu_1 p_{10} \simeq a$, the minimum indecomposability of the aggregates is $2a\{1-\cos[\pi/(N+1)]\}$; thus the sufficient conditions for the near-complete decomposability of $\mathbf{Q}(N, 2)$ become

$$\varepsilon_1/a < 1 - \cos[\pi/(N+1)], \tag{4.21}$$

and this specifies an upper bound for ε_1/a that is the maximum ratio of coupling between and within aggregates.

Table 4.1 gives the value of $\{1-\cos[\pi/(N+1)]\}$ for some typical values of N.

This table shows how the condition (4.21) becomes rapidly stringent when the population N increases, especially if one remembers (Chapter III) that this condition guarantees an approximation of less than $\frac{1}{2}$ only. This can be explained by the fact that each aggregate $\mathbf{Q}(n, 1)$ has here the same structure

TABLE 4.1

N	$1-\cos[\pi/(N+1)]$
1	1
5	0.13
10	0.40×10^{-1}
100	0.48×10^{-3}
1000	0.49×10^{-5}
10^k	$\approx 10^{-2k}\pi^2/2$

as a single queue with server R_0, input station R_1, and finite population n. Queues are known to converge rather slowly toward their equilibrium; actually, the convergence rate is that of a geometrical series of parameter $1/\max_I |\lambda^*(2_I)|$, which is here equal to $1/(1-2a\{1-\cos[\pi/(N+1)]\})$. This is true for all levels of aggregation in queueing networks. As the number $\binom{L+N}{L}$ of states increases more rapidly with L than with N, and since condition (4.20) is independent of L, the analysis of queueing networks by partition and aggregation appears therefore all the more advantageous when L is large as compared with N. We shall come back to this point later when we apply this approach to models of computing systems.

4.5 Central-Server Model

In the following chapters, we shall frequently be concerned with a particular case of the Jacksonian queueing network model $\mathbf{Q}(N, L)$. In this particular model, referred to by Buzen (1971a) as the *central-server model*, we have, taking R_0 as the central server,

$$\forall k \neq 0: \quad p_{kk} = 0,$$

and

$$\forall k, m \neq 0: \quad p_{k0} = 1, \qquad p_{km} = 0.$$

In this case, in conditions (4.20), ω_l, A_l, and B_l become

$$\omega_l = \max_{\substack{i_0, i_{l+1}, \ldots, i_L \\ \sum_k i_k \leqslant N}} \left[\sum_{k=l+1}^{L} \kappa(i_k)\,\mu_k + \kappa(i_0)\,\mu_0 \sum_{m=l+1}^{L} p_{0m} \right], \tag{4.22}$$

$$A_l = \mu_0\, p_{0l}, \tag{4.23}$$

$$B_l = \mu_l. \tag{4.24}$$

If $N \geqslant L+1$, then ω_l is more simply defined by Corollary 4.1:

$$\omega_l = \sum_{k=l+1}^{L} \mu_k + \mu_0 \sum_{m=l+1}^{L} p_{0m}. \tag{4.25}$$

This central-server model is illustrated by Fig. 4.4.

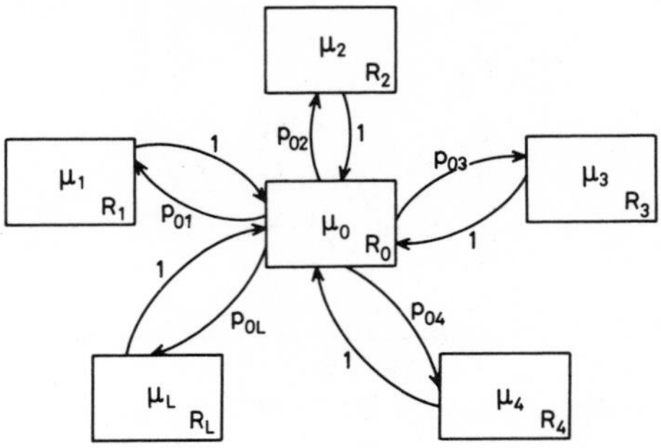

FIGURE 4.4 Central-server model.

CHAPTER V

Hierarchy of Aggregate Resources

The object of this chapter is to describe how a queueing network $\mathbf{Q}(N, L)$, $N > 0$, $L > 1$, that is $(L-1)$-level nearly completely decomposable can be analyzed by an aggregation technique. We therefore assume that there exists an ordering of the resources $R_0, R_1, \ldots, R_L$ such that the service rates μ_l and the customer stochastic behavior $\{p_{lk}\}$, $0 \leqslant l$, $k \leqslant L$, obey the $L-1$ inequalities (4.20). In this case, $L-1$ levels of aggregation may be distinguished, each level being associated with a resource R_l; the network is at each level l nearly completely decomposable into $N+1$ aggregates represented by matrices $\mathbf{Q}(n, l)$, $n = N, \ldots, 0$, $l = L-1, \ldots, 1$. In this chapter, our specific aim is to define the aggregative variables that have to represent the internal equilibrium states reached by these aggregates; in the spirit of the multilevel aggregation approach outlined in Section 1.5, these variables must be such that at each level they can be evaluated in terms of the aggregative variables of the lower levels. The advantages of this method of analysis will be discussed in the next chapter.

5.1 Decomposition into Levels of Aggregation

A matrix $\mathbf{Q}(n, l)$, $n = 0, \ldots, N$, $l = 1, \ldots, L$, defines an aggregate in which n customers compete for and use the $l+1$ resources $R_0, R_1, \ldots, R_l$. We assume that $\mathbf{Q}(N, L)$ is $(L-1)$-level nearly completely decomposable; hence, for each level l there exists a time period (cf. Section 1.5) that starts at a certain time $T_{0,l}$ and during which the aggregates of level l evolve almost independently of one another and of the remainder of the system toward a statistical equilibrium reached at some later time $T_{1,l}$. During this period the statistical equilibrium states that had been reached by the aggregates $\mathbf{Q}(m, l-1)$, $m = 0, \ldots, n$, at some time $T_{1,l-1}$ are approximately preserved. The property of near-complete decomposability guarantees that ε_l, the maximum degree of coupling between aggregates, is small enough so that $T_{1,l-1} < T_{0,l}$.

For the time period of interest for the analysis of the behavior of aggregates $\mathbf{Q}(n,l)$, we can therefore assume that aggregates $\mathbf{Q}(m,l-1)$ are in statistical equilibrium, viz., that for any n, $n = 0, \ldots, N$, the distribution of customers among $R_0, \ldots, R_{l-1}$ is approximately stationary. As a consequence of this, we can regard the time behavior of $\mathbf{Q}(n,l)$ as a convergence toward equilibrium of the customer distribution between a resource R_l on the one hand, and an *aggregate resource* on the other, which consists of the resources $R_0, \ldots, R_{l-1}$, among which a "quasi-stationary" statistical equilibrium has already established itself. To this purpose, we shall associate with each aggregate $\mathbf{Q}(n,l)$ a queueing system denoted $\mathcal{M}_l(n_l)$, and schematically represented in Fig. 5.1.

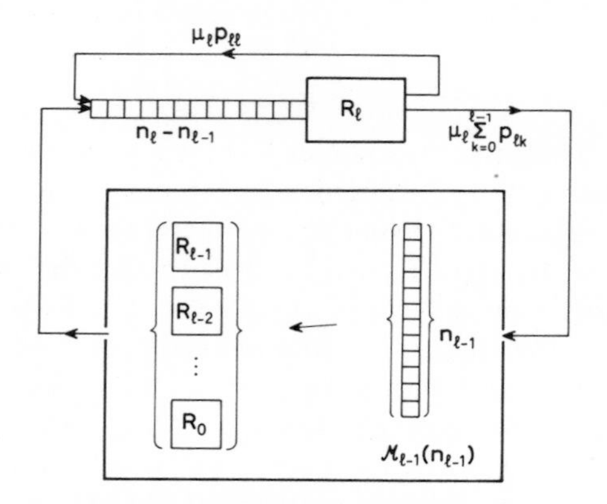

FIGURE 5.1 Queueing system $\mathcal{M}_l(n_l)$.

In this system, a population of n_l cycling customers requests alternatively R_l and the aggregate resource. We shall say that the system $\mathcal{M}_l(n_l)$ is in state $E_l(n_{l-1} \mid n_l)$, $n_{l-1} = 0, \ldots, n_l$, whenever n_{l-1} among the n_l customers are either waiting for or being serviced by that aggregate resource, i.e., whenever n_{l-1} among n_l customers are distributed over the resources $R_{l-1}, \ldots, R_0$, and $n_l - n_{l-1}$ customers are either waiting for or being serviced by the resource R_l.

5.2 Level Analysis

The time behavior of $\mathcal{M}_l(n_l)$ may be analyzed in the following way. For the time period $[T_{0,l}, T_{1,l}]$, the three following assumptions can be made, as a result of the near-complete decomposability of $\mathbf{Q}(N,L)$:

5.2.1 We can disregard all interactions between $\mathcal{M}_l(n_l)$ on the one hand and $R_{l+1}, \ldots, R_L$ on the other; in other words, we may regard $\mathcal{M}_l(n_l)$ as a closed system that no customer can leave or enter. The finite population n_l may thus be considered as remaining constant and—since all customers have the same stochastic behavior $\{p_{jk}\}$, $j,k = 0,\ldots,l$—as remaining identical between $T_{0,l}$ and $T_{1,l}$.

5.2.2 Just as $\mathbf{Q}(n,l)$, so can $\mathcal{M}_l(n_l)$ be regarded as an irreducible system in which each state $E_l(k \mid n_l)$, $k = 0,\ldots,n_l$, can be reached from any other state $E_l(j \mid n_l)$ within a finite time period. This property is in fact enjoyed by most queueing networks with finite population, with resources of nonzero service rates, and with a customer behavior defined by a connected graph of edges p_{jk}, $j,k = 0,\ldots,l$. Thereby $\mathcal{M}_l(n_l)$ enjoys the property of ergodicity: there exists a stationary distribution of the probabilities, say,

$$\pi_l(n_{l-1} \mid n_l) = \Pr\{\mathcal{M}_l(n_l) \text{ is in state } E_l(n_{l-1} \mid n_l)\}, \qquad n_{l-1} = 0,\ldots,n_l,$$

which is independent of time and of the initial state of $\mathcal{M}_l(n_l)$.

5.2.3 Just as $\mathbf{Q}(n,l)$ is itself nearly completely decomposable into subaggregates $\mathbf{Q}(m,l-1)$, $m = 0,\ldots,n$, $\mathcal{M}_l(n_l)$ is nearly completely decomposable into subsystems $\mathcal{M}_{l-1}(n_{l-1})$, $n_{l-1} = 0,\ldots,n_l$. By virtue of our decomposition scheme, *the aggregate resource of $\mathcal{M}_l(n_l)$ is nothing but the queueing system $\mathcal{M}_{l-1}(n_{l-1})$ whenever $\mathcal{M}_l(n_l)$ is in state $E_l(n_{l-1} \mid n_l)$*; and the quasi-stationary state reached by this aggregate resource $\mathcal{M}_{l-1}(n_{l-1})$ at some time $T_{1,l-1} < T_{0,l}$ is defined by the stationary distribution

$$\{\pi_{l-1}(n_{l-2} \mid n_{l-1})\}, \qquad n_{l-2} = 0,\ldots,n_{l-1}.$$

With these assumptions, the distribution $\{\pi_l(n_{l-1} \mid n_l)\}$ may be obtained as the distribution of the congestion n_{l-1} in a $M \mid M \mid 1 \mid n_l$ queueing system $\mathcal{M}_l(n_l)$ with finite population n_l. The probability that a state transition $E_l(n_{l-1} \mid n_l) \to E(n_{l-1}+1 \mid n_l)$, $n_{l-1} = 0,\ldots,n_l-1$, occurs in this system during a time interval $(t,\, t+h_l)$, $t < T_{0,l+1}$, is equal to

$$h_l \mu_l \sum_{k=0}^{l-1} p_{lk} + O(h_l), \tag{5.1}$$

where $O(h_l)$ stands for the probability of a combined occurrence between t and $t+h_l$ of two or more services completed at resource R_l and applying to $R_0,\ldots,R_{l-1}$, and of one or more services completed at $R_0,\ldots,R_{l-1}$ and applying to R_l. This probability is obviously of the order of magnitude of h_l^2 and can be neglected if h_l is taken sufficiently small with respect to the time scale of interactions between these resources. Likewise, the probability of a transition $E_l(n_{l-1} \mid n_l) \to E_l(n_{l-1}-1 \mid n_l)$, $n_{l-1} = 1,\ldots,n_l$, during a time interval $(t,\, t+h_l)$, $t > T_{1,l-1}$, is equal to

$$h_l \psi_{l-1,l}(n_{l-1}) + O(h_l), \tag{5.2}$$

where $h_l \psi_{l-1,l}(n_{l-1})$ is used to denote the probability that during any such interval, one customer completes service at one of the resources $R_0, \ldots, R_{l-1}$ of $\mathcal{M}_{l-1}(n_{l-1})$ *and* applies to resource R_l. In fact, $\psi_{l-1,l}(n_{l-1})$ is one of the aggregative variables we shall use to represent at level l the stationary state that has been reached at time $T_{1,l-1}$ by $\mathcal{M}_{l-1}(n_{l-1})$. Later we shall examine how these aggregative variables can be evaluated.

The balance equations that define the statistical equilibrium of such a queueing system $\mathcal{M}_l(n_l)$, $n_{l-1} = 1, \ldots, n_l - 1$ [see, e.g., Feller (1968, p. 460 ff)] reduce to

$$\begin{aligned}
&\left[\psi_{l-1,l}(n_{l-1}) + \mu_l \sum_{k=0}^{l-1} p_{lk}\right] \pi_l(n_{l-1} \mid n_l) \\
&\qquad = \pi_l(n_{l-1} - 1 \mid n_l)\, \mu_l \sum_{k=0}^{l-1} p_{lk} + \pi_l(n_{l-1} + 1 \mid n_l)\, \psi_{l-1,l}(n_{l-1} + 1), \\
&\left[\mu_l \sum_{k=0}^{l-1} p_{lk}\right] \pi_l(0 \mid n_l) = \pi_l(1 \mid n_l)\, \psi_{l-1,l}(1), \\
&\psi_{l-1,l}(n_l)\, \pi_l(n_l \mid n_l) = \pi_l(n_l - 1 \mid n_l)\, \mu_l \sum_{k=0}^{l-1} p_{lk}.
\end{aligned} \tag{5.3}$$

These equations express that for each state the probability of departure is equal to the probability of arrival. They yield closed-form expressions for the probabilities $\{\pi_l(n_{l-1} \mid n_l)\}$ as a function of the parameters μ_l and p_{lk}, and the aggregative variables $\psi_{l-1,l}(n_{l-1})$:

$$\pi_l(n_{l-1} \mid n_l) = \frac{(\mu_l \sum_{k=0}^{l-1} p_{lk})^{n_{l-1}}}{\prod_{k=1}^{n_{l-1}} \psi_{l-1,l}(k)}\, \pi_l(0 \mid n_l), \qquad n_{l-1} = 1, \ldots, n_l, \qquad n_l = 1, \ldots, N. \tag{5.4}$$

$\pi_l(0 \mid n_l)$ is obtained with the additional condition that the probabilities $\pi_l(n_{l-1} \mid n_l)$, $n_{l-1} = 0, \ldots, n_l$ must add to unity. We are left with the evaluation of the aggregative variables $\psi_{l-1,l}(n_{l-1})$, or more generally, because we shall need them, with the evaluation of all the probabilities $\psi_{l,k}(n_l)$, $k > l$, $l = 0, \ldots, L-1$ that a customer completes service at the aggregate resource $\mathcal{M}_l(n_l)$ and applies to some upper resource R_k, during any small time interval after time $T_{1,l}$ when $\mathcal{M}_l(n_l)$ is in equilibrium. The level-by-level analysis takes place here. The probability of completing service in $\mathcal{M}_l(n_l)$ is equal to the probability of completing service either at R_l or at one of the aggregate resources $\mathcal{M}_{l-1}(n_{l-1})$, $n_{l-1} = 1, \ldots, n_l$. However, after time $T_{1,l}$ the aggregate resource of $\mathcal{M}_l(n_l)$ is $\mathcal{M}_{l-1}(n_{l-1})$ with steady-state probability $\pi_l(n_{l-1} \mid n_l)$; hence $\psi_{l,k}(n_l)$ obeys the recurrence relations

$$\psi_{l,k}(n_l) = [1 - \pi_l(n_l \mid n_l)]\, \mu_l p_{lk} + \sum_{n_{l-1}=1}^{n_l} \pi_l(n_{l-1} \mid n_l)\, \psi_{l-1,k}(n_{l-1}), \qquad k > l, \qquad l = 1, \ldots, L-1, \qquad n_l = 1, \ldots, N. \tag{5.5}$$

These relations yield $\psi_{l,k}(n_l)$ as a function of $\psi_{l-1,k}(n_{l-1})$. For $l=0$ and $k>0$, expression $h_0\psi_{0,k}(n_0)+O(h_0)$ is simply the probability that a customer completes service at R_0 and applies to R_k. Thus, for all $n_0 \neq 0$ we have

$$\psi_{0,k}(n_0) = \mu_0 p_{0k}. \tag{5.6}$$

To sum up, the level-by-level procedure to obtain the probability distributions

$$\pi_l(n_{l-1}\,|\,n_l) \qquad \text{for} \quad n_{l-1} = 0,\ldots,n_l, \quad n_l = 1,\ldots,N, \quad l = 1,\ldots,L,$$

is the following:

(1) For the lowest aggregation level $l=1$, which is the first one to reach equilibrium, obtain the distribution $\{\pi_1(n_0\,|\,n_1)\}$ by (5.4) with $\psi_{0,k}(n_0) = \mu_0 p_{0k}$. Obtain the aggregative variables $\psi_{1,k}(n_1)$ by (5.5) for all $k=2,\ldots,L$.

(2) Successively, for each level $l=2,\ldots$m L, obtain the distributions $\pi_l(n_{l-1}\,|\,n_l)$ by (5.4) as a function of the $\psi_{l-1,l}(n_{l-1})$. Obtain by (5.5) the variables $\psi_{l,k}(n_l)$ required for the upper levels $k=l+1,\ldots,L$ as a function of $\psi_{l-1,k}(n_{l-1})$.

5.2.4 ***Remark*** The recurrence relations (5.5) can be expanded. Factoring out the transfer probabilities $\mu_j p_{jk}$, $j=0,\ldots,l$, we obtain

$$\psi_{l,k}(n_l) = \sum_{j=0}^{j=l} A_l(j,n_l)\,\mu_j p_{jk}, \tag{5.7}$$

where the coefficients $A_l(j,n_l)$, $j \leqslant l$, are

$$A_l(l,n_l) = 1 - \pi_l(n_l\,|\,n_l), \qquad \text{for} \quad j = l, \tag{5.8}$$

$$A_l(j,n_l) = \sum_{n_{l-1}=1}^{n_l} \cdots \sum_{n_{j+1}=1}^{n_{j+2}} \sum_{n_j=1}^{n_{j+1}} \pi_l(n_{l-1}\,|\,n_l)\cdots\pi_{j+2}(n_{j+1}\,|\,n_{j+2})$$
$$\times\, \pi_{j+1}(n_j\,|\,n_{j+1})[1-\pi_j(n_j\,|\,n_j)], \qquad \text{for} \quad j = 0,\ldots,l-1, \tag{5.9}$$

taking the convention that $\pi_0(n_0\,|\,n_0) = 0$ for $n_0 \neq 0$.

As can be seen from (5.9), the coefficients $A_l(j,n_l)$ also obey a recurrence relation

$$A_l(j,n_l) = \sum_{n_{l-1}=1}^{n_l} \pi_l(n_{l-1}\,|\,n_l)\,A_{l-1}(j\,|\,n_{l-1}), \qquad j = 0,\ldots,l-1, \tag{5.10}$$

with $A_0(0,n_0) = 1$.

5.2.5 ***Remark*** The probability $\psi_{l,k}(n_l)$ is but a particular case of an aggregative variable representative of the quasi-stationary state reached at level l. Suppose for instance that we are interested in the service rate of the system

$\mathscr{M}_l(n_l)$, say $\sigma_l(n_l)$, defined as the mean number of requests serviced by that "machine" per time unit. This service rate obeys also recurrence relations similar to (5.5):

$$\sigma_l(n_l) = \mu_l[1-\pi_l(n_l \mid n_l)] + \sum_{n_{l-1}=1}^{n_l} \pi_l(n_{l-1} \mid n_l)\, \sigma_{l-1}(n_{l-1}), \tag{5.11}$$
$$l = 1, \ldots, L, \qquad n_l = 1, \ldots, N,$$

and $\sigma_0(n_0) = \mu_0$ for $n_0 \neq 0$.

5.2.6 ***Remark*** For the central-server model (cf. Section 4.5), $\psi_{l,k}(n_l)$ reduces to only the second term of (5.5).

5.3 Interlevel Relationship

So far, each aggregate $\mathscr{M}_l(n_l)$, $l = 1, \ldots, L-1$, has been analyzed as a closed system inaccessible to entry into or exit from the upper levels, and the equilibrium distributions that have been obtained are conditioned by the fixed number n_l of customers supposed to be permanently cycling in each aggregate $\mathscr{M}_l(n_l)$. These conditional internal equilibria are what Simon and Ando call the short-term equilibria, which are successively attained at each level of aggregation. We have seen in Section 1.5 how the long-term equilibrium probabilities for each lower level of aggregation could be derived from the long-term equilibrium at the uppermost level. If $a_l(n_{l-1})$, $n_{l-1} = 0, \ldots, N$, $l = 1, \ldots, \ldots, L$, is the *unconditional* long-term equilibrium probability of n_{l-1} customers being in service or in queue at the *a*ggregate resource $R_0, \ldots, R_{l-1}$, and $s_l(i_l)$, $i_l = 0, \ldots, N$ is the *unconditional* long-term equilibrium probability of i_l customers being in service or in queue at the *s*ingle resource R_l, relations similar to (1.37) yield

$$a_L(n_{L-1}) = \pi_L(n_{L-1} \mid N), \qquad s_L(i_L) = \pi_L(N-i_L \mid N), \quad n_{L-1}, i_L = 0, \ldots, N, \tag{5.12}$$

and for each level $l = L-1, \ldots, 1$,

$$a_l(n_{l-1}) = \sum_{n_l=n_{l-1}}^{N} a_{l+1}(n_l)\, \pi_l(n_{l-1} \mid n_l), \tag{5.13}$$

$$s_l(i_l) = \sum_{n_l=i_l}^{N} a_{l+1}(n_l)\, \pi_l(n_l - i_l \mid n_l). \tag{5.14}$$

At level 1, whose only constituents are resources R_1 and R_0, we have

$$s_0(i_0) \equiv a_1(i_0), \qquad i_0 = 0, \ldots, N. \tag{5.15}$$

These relations simply express the fact that the probability of having a population of n_l customers cycling at level l is equal to the probability of being in one of the states $E_{l+1}(n_l \mid n_{l+1})$, $n_{l+1} = n_l, \ldots, N$ at level $l+1$.

The long-term fraction of time a resource R_l, $l = 0, \ldots, L$, is busy equals $1 - s_l(0)$. The *mean response time* W_l of resource R_l, defined as the mean time spent by a customer in queue and in service at resource R_l, may be easily deduced using Little's (1961) formula

$$L = \lambda W, \tag{5.16}$$

where L is the mean number of units in a queueing system, $1/\lambda$ the mean time between consecutive arrivals to this system, and W the mean time spent by a unit in the system. Little's formula applied to the queue of resource R_l yields

$$W_l = \{[1 - s_l(0)]\,\mu_l\}^{-1} \sum_{i_l=1}^{N} i_l s_l(i_l), \qquad l = 0, \ldots, L, \tag{5.17}$$

since in the statistical equilibrium the input rate to queue R_l equals its output rate $[1 - s_l(0)]\,\mu_l$.

Moreover, if the uppermost level resource R_L is the model of customer arrivals at a rate μ_L obeying Eq. (4.2),

$$\mu_L = \lambda \times i_L, \qquad i_L = 0, \ldots, N,$$

using Little's formula one can deduce from the probabilities $a_L(n_{L-1})$, $n_{L-1} = 0, \ldots, N$, the mean time spent by a customer in a system $R_0, \ldots, R_{L-1}$:

$$W = \left[\lambda \sum_{i_L=1}^{N} i_L s_L(i_L)\right]^{-1} \sum_{n_{L-1}=1}^{N} n_{L-1} a_L(n_{L-1}). \tag{5.18}$$

5.4 Conclusions

The above approach is an example of how the time analysis of a system may be broken up into distinct stages at each of which only a subspace of the system state space needs to be taken into consideration. In our analysis, we have substituted the closed multiqueue system $\mathbf{Q}(N, L)$ whose state space is of size $\binom{N+L}{L}$, with $L \times N$ single-server queueing systems $\mathcal{M}_l(n_l)$, $l = 1, \ldots, L$, $n_l = 1, \ldots, N$, each having $n_l + 1$ distinct states. Each system $\mathcal{M}_l(n_l)$ is in fact an equivalent representation for the $\binom{L-(l+1)+n_l}{L-(l+1)}$ subsystems represented by the principal submatrices $\mathbf{Q}'(i_{l+1}, \ldots, i_L)$, $\sum_{k=l+1}^{N} i_k = N - n_l$, of $\mathbf{Q}(N, L)$, which can all be classified into a same equivalence class $\mathbf{Q}(n_l, l)$ (see the Proof to Theorem 4.1).

This kind of state space partitioning will, of course, also be instrumental in the analysis of larger and more complex models than $\mathbf{Q}(N, L)$, which is the object of the next chapter.

CHAPTER VI

Queueing-Network Analysis

In Chapters IV and V we restricted our attention to the rather simple model of queueing network defined in Section 4.1. One reason for this was to keep the presentation as clear as possible; the other, more important reason was that we could, in this simple model, obtain a manageable expression of the aggregate minimum indecomposability. With more complex models it would have been more difficult, if not impossible, to obtain such an expression.

In this chapter we first introduce some generalizations of our basic model that can still be analyzed by the hierarchical approach of resource aggregation presented in Chapter V. Then we take a more general standpoint and discuss the major practical advantages that may be expected from any model based on the principles of near-complete decomposability and variable aggregation.

6.1 State Dependency

A first possible generalization is to allow, as already done in Jackson's (1963) model, the service rates of any resource R_l, $l = 0, \ldots, L$, to be an arbitrary nonnegative function $\mu_l(i_l)$ of the number i_l of customers currently in queue or in service at this resource [$\forall l$: $\mu_l(0) = 0$]. This generalization is useful to model resources like storage devices, for example, the behavior of which varies with their occupancy; we shall in fact make use of this generalization in Chapter IX.

Assuming that the transfers between resources remain governed by the same probabilities $\{p_{kl}\}$, sufficient conditions under which this generalized model of queueing network is nearly completely decomposable may be obtained as follows. The maximum degree of coupling ε_l, introduced in Theorem 4.1, is now at most equal to

$$\omega_l = \max_{\substack{i_0, \ldots, i_L \\ \sum_k i_k \leqslant N}} \left[\sum_{k=l+1}^{L} \kappa(i_k)\, \mu_k(i_k) \sum_{m=0}^{l} p_{km} + \sum_{k=0}^{l} \kappa(i_k)\, \mu_k(i_k) \sum_{m=l+1}^{L} p_{km} \right], \quad (6.1)$$

at most N functions $\kappa(i_k)$ being not zero. Using the same approach as in Section 4.3, a lower bound for the indecomposability of the aggregates at level l can be easily deduced from the probabilities (4.14) and (4.15):

$$A_l + B_l - 2(A_l B_l)^{1/2} \cos[\pi/(N+1)], \tag{6.2}$$

where

$$A_l = \min_{\substack{i_0, \ldots, i_{l-1} \\ 1 \leqslant \sum_k i_k \leqslant N}} \left[\sum_{k=0}^{l-1} \kappa(i_k)\, \mu_k(i_k)\, p_{kl} \right], \tag{6.3}$$

$$B_l = \min_{1 \leqslant i_l \leqslant N} \left[\mu_l(i_l) \sum_{k=0}^{l-1} p_{lk} \right]. \tag{6.4}$$

The sufficient conditions for this generalized model to be $(L-1)$-level nearly completely decomposable are thus obtained by substituting ω_l, A_l, and B_l with the above values in the inequalities (4.20).

The interaction rate of the aggregate $\mathscr{M}_l(n_l)$, $n_l = 1, \ldots, N$, upon resource R_k, $k > l$, may then be obtained as

$$\begin{aligned} \psi_{l,k}(n_l) = & \sum_{n_{l-1}=0}^{n_l - 1} \pi_l(n_{l-1} \mid n_l)\, p_{lk}\, \mu_l(n_l - n_{l-1}) \\ & + \sum_{n_{l-1}=1}^{n_l} \pi_l(n_{l-1} \mid n_l)\, \psi_{l-1,k}(n_{l-1}). \end{aligned} \tag{6.5}$$

The balance equations of this aggregate are similar to Eqs. (5.4), and yield the stationary distribution

$$\pi_l(n_{l-1} \mid n_l) = \frac{(\sum_{k=0}^{l-1} p_{lk})^{n_{l-1}} \prod_{k=0}^{n_{l-1}-1} \mu_l(n_l - k)}{\prod_{k=1}^{n_{l-1}} \psi_{l-1,l}(k)}\, \pi_l(0 \mid n_l), \tag{6.6}$$

$$n_{l-1} = 1, \ldots, n_l, \qquad n_l = 1, \ldots, N.$$

$\pi_l(0 \mid n_l)$ is obtained by the additional condition that the sum of these probabilities must be equal to unity. Expressions (5.13)–(5.15) yield the unconditional long-term equilibrium probabilities $s_l(i_l)$, $i_l = 0, \ldots, N$, of i_l customers being in queue or in service at resource R_l, $l = 0, \ldots, L$.

Gordon and Newell (1967) studied the particular case of the above generalization in which $\mu_l(i_l)$ is a nondecreasing function of i_l given by

$$\mu_l(i_l) = \alpha_l(i_l)\, \mu_l, \qquad l = 0, \ldots, L,$$

where

$$\alpha_l(i_l) = \begin{cases} i_l, & \text{if } i_l \leqslant v_l, \\ v_l, & \text{if } i_l \geqslant v_l. \end{cases}$$

Each level of such a system consists in fact of v_l parallel exponential servers, each with mean service rate μ_l. Replacing $\mu_l(i_l)$ in (6.3) and (6.4) by the above function shows clearly that A_l and B_l keep their values (4.18) and (4.19) of

Section 4.3. More generally, *when the service rate at each level is a function of the congestion i_l, the aggregate indecomposability lower bounds remain those of a simple model* $\mathbf{Q}(N, L)$ *with each resource being assigned the minimum service rate*; however, ω_l must always be calculated by (6.1).

By a similar generalization, state-dependent transfer probabilities—let us denote them $p_{lk}(i_l, i_k)$—that depend on the congestion at the stage of departure and/or the stage of arrival may also be coped with. The above expressions for ω_l, A_l, B_l, and $\{\pi_l(n_{l-1} \mid n_l)\}$ are easily rewritten in this case. Moreover, the aggregative probability $\psi_{l,k}(n_l)$ also becomes dependent on the state of the stage of arrival R_k, and the following aggregative variables are needed:

$$\psi_{l,k}(n_l, i_k) = \sum_{n_{l-1}=0}^{n_l-1} \pi_l(n_{l-1} \mid n_l)\, \mu_l(n_l - n_{l-1})\, p_{lk}(n_l - n_{l-1}, i_k) + \sum_{n_{l-1}=0}^{n_l} \pi_l(n_{l-1} \mid n_l)\, \psi_{l-1,k}(n_{l-1}), \qquad i_k = 0, \ldots, N - n_l. \tag{6.7}$$

6.1.1 ***Remark*** Relations (6.7) show that, when transfer probabilities are dependent on the state i_k of the arrival stage, $(N+1)(N+2)/2$ variables $\psi_{l,k}(n_l, i_k)$, $i_k = N - n_l$, are required for a given pair (l, k), while only $N+1$ variables $\psi_{l,k}(n_l)$, $n_l = 0, \ldots, N$, are needed in the simple model. This increase may seem at first sight disproportionate since the number of distinct system states remains identical and equal to $\binom{L+N}{L}$. But the complexity of a model is a function not only of its number of distinct states, but also of the various interactions between states; the number of these interactions, i.e., of the transfer probabilities, is here multiplied by $N+1$.

6.1.2 ***Remark*** Jacksonian models (Jackson, 1963) do not take into account state-dependent transfer probabilities. On the other hand, they can cope with networks in which the total number of customers present in the system may, in principle, vary from zero to infinity, the customer arrival process being taken as an arbitrary function of this total number. As already suggested in Section 4.1, the model $\mathbf{Q}(N, L)$ and thus the hierarchical approach of Chapter V can, by means of an additional stage of service, cope with a population of variable size, but this population cannot exceed a fixed finite value; in practice, however, this cannot be a serious limitation.

6.2 Arbitrary Service Time Distributions

6.2.1 It is a well-known property that the superposition of *independent* renewal processes leads to a compound process that rapidly becomes similar to a Poisson process when the number of independent component processes increases. More precisely, consider a number S of independent sources at each of which events occur from time to time; suppose that the intervals between successive events at any one source are independent random variables, so that each source output is a renewal process. When the outputs of the sources are combined into one pooled output, the distribution of the intervals between successive events in this pooled process approaches an exponential function as S increases, whatever the distributions of the intervals in the component processes are (Khintchine, 1955; Cox and Smith, 1954; Ten Hoopen and Reuver, 1966). This property has some useful consequences in nearly completely decomposable systems.

Aggregates, by definition, behave approximately independently of one another; it follows that their output processes, i.e., the time sequences of departures from each aggregate, are approximately mutually independent processes. The above property is therefore applicable: the pooled output of a sufficiently large number of aggregates is approximately Poisson, regardless of the type of distribution of each component output. Hence, an aggregative random variable that models the superimposed behavior of several aggregates may justifiably be considered exponentially distributed even though the individual aggregate output processes are non-Poisson.

This superposition property has already been applied in the hierarchical model analyzed by Courtois and Georges (1970), but as far as we know no attention has yet been given to it in more general studies related to techniques of variable aggregation. In this section we study only one aspect of it: our hierarchical model of Chapter V can be used to analyze queueing networks in which, instead of being exponentially distributed, the service times of certain resources have *general* distribution functions.

Considering a queueing network, in all other respects similar to our basic model $\mathbf{Q}(N, L)$, we shall use $B_l(x)$ to denote the general service time distribution function of resource R_l, $l = 0, \ldots, L$, and β_l to denote the mean of this distribution

$$\beta_l = \int_0^\infty x \, dB_l(x). \tag{6.8}$$

6.2.2 The first step is to prove that

Theorem 6.1 *In a queueing network with arbitrary service time distributions, the probability of a customer completing service in an aggregate $\mathcal{M}_l(n_l)$ and applying to resource R_k, $k > l$, during any small time interval $(t, t+h_l]$ after time $T_{1,l}$ when $\mathcal{M}_l(n_l)$ is in equilibrium, is equal to $h_l \psi_{l,k}(n_l)$, with $\psi_{l,k}(n_l)$, $1 \leqslant l < k \leqslant L$, $0 \leqslant n_l \leqslant N$, defined by the recurrent relations*

$$\psi_{l,k}(n_l) = [1 - \pi_l(n_l \mid n_l)]\beta_l^{-1} p_{lk} + \sum_{n_{l-1}=1}^{n_l} \pi(n_{l-1} \mid n_l) \psi_{l-1,k}(n_{l-1}), \tag{6.9}$$

where $\psi_{0,k}(n_0) = \beta_0^{-1} p_{0k}$ *for* $n_0 \neq 0$, *and* $\{\pi_l(n_{l-1} | n_l)\}$, $n_{l-1} = 0, \ldots, n_l$, *is the stationary distribution of the congestion in* $\mathcal{M}_l(n_l)$.

PROOF We define $m_l(t)$, $t \geqslant 0$, as the number of services completed at time t by resource R_l, and we consider for any time t the function $d_l(t)$ defined by

$$d_l(t) = \lim_{h \to 0} \{h^{-1} [m_l(t+h) - m_l(t)]\}. \tag{6.10}$$

This is the *renewal density* (Cox, 1962) of the renewal process, which consists of the time sequence of the service completions at resource R_l. $hd_l(t)$ is asymptotically also the probability of a service completion by resource R_l during the small interval $(t, t+h]$. If we let $\tau_0 = 0, \tau_1, \tau_2, \ldots$ be the sequence of instants at which a service is completed by R_l, a central result of the renewal theory [Blackwell's theorem; see, e.g., Takacs (1962, p. 225)] states that for every $h > 0$

$$\lim_{t \to \infty} \{h^{-1} [m_l(t+h) - m_l(t)]\} = 1/E\{(\tau_n - \tau_{n-1})\}, \tag{6.11}$$

provided that the time interval $(\tau_n - \tau_{n-1})$ between two successive service completions is a random variable with nonlattice distribution, and that its mean $E\{(\tau_n - \tau_{n-1})\}$ is finite. Assuming these mild restrictions satisfied, from (6.10) and (6.11) we obtain

$$\lim_{t \to \infty} d_l(t) = 1/E\{(\tau_n - \tau_{n-1})\}. \tag{6.12}$$

Letting θ_n be the amount of time during which R_l is idle within the interval (τ_{n-1}, τ_n), we have

$$E\{(\tau_n - \tau_{n-1})\} = E\{\theta_n\} + \beta_l. \tag{6.13}$$

By definition, we also have

$$\lim_{t \to \infty} \Pr\{R_l \text{ is idle at time } t\} = E\{\theta_n\}/E\{(\tau_n - \tau_{n-1})\}.$$

From (6.12) and (6.13), we conclude that

$$\lim_{t \to \infty} d_l(t) = \beta_l^{-1} (1 - \lim_{t \to \infty} \Pr\{R_l \text{ is idle at time } t\}). \tag{6.14}$$

Let us now consider the aggregate $\mathcal{M}_l(n_l)$ after time $T_{1,l}$ when it is in equilibrium. We again assume that conditions (irreducibility) are satisfied for the stationary distribution $\{\pi_l(n_{l-1} | n_l)\}$, $n_{l-1} = 0, \ldots, n_l$, to exist. A service is completed in $\mathcal{M}_l(n_l)$ either by one of the lower level aggregates $\mathcal{M}_{l-1}(n_{l-1})$ or by resource R_l. By definition of $\pi_l(n_l | n_l)$, we now have

$$\lim_{t \to \infty} d_l(t) = \beta_l^{-1} [1 - \pi(n_l | n_l)]. \tag{6.15}$$

Moreover, for $t > T_{1,l}$, the queue at R_l is in equilibrium and $d_l(t)$ is near its limiting value; the larger the dispersion of $B_l(t)$ is, the faster this limiting value is approached (Cox, 1962, p. 55). Thus, the probability of a service being completed by R_l during any time interval $(t, t+h_l]$, $t > T_{1,l}$, can be validly taken equal to $\beta_l^{-1}[1-\pi_l(n_l \mid n_l)]$. Then, using $h_l \psi_{l,k}(n_l)$ to denote the probability of a service being completed by aggregate $\mathscr{M}_l(n_l)$ and applying to resource R_k, $k > l$ during any small time interval $(t, t+h_l]$, $t > T_{1,l}$, we obtain that $\psi_{l,k}(n_l)$ is defined by the same recurrence relations (5.5), in which μ_l is simply replaced by β_l^{-1}. This completes the proof.

6.2.3 Theorem 6.1 has two consequences. First it shows that in the definition of the flow density of customers applying to a resource R_k from a lower level aggregate resource already in equilibrium, the distribution functions of the service times of the lower level component resources enter only with their means. Stated otherwise, once an aggregate is in equilibrium, the higher moments of the service distributions of its component resources have little influence on its output traffic. This situation is comparable to the equilibrium state of a semi-Markov process associated with an ergodic discrete Markov chain; the limiting state probabilities of the semi-Markov process are functions of the equilibrium state probabilities of the Markov chain and of the means only of the distribution functions of the time intervals between successive transitions.

A second, more practical consequence of Theorem 6.1 is that, after time $T_{1,l}$ when aggregates of level l are in equilibrium, their output processes are approximately *Poisson processes* since $\psi_{l,k}(n_l)$, $k > l$, is constant for any time interval $(t, t+h_l]$, $t > T_{1,l}$. This property holds true whatever the service time distributions of the lower level resources $R_0, \ldots, R_l$ are. Therefore, from this level on, each queueing system $\mathscr{M}_{l+1}(n_{l+1})$, $n_{l+1} = 1, \ldots, N$, has approximately the same transient time behavior during the period $(T_{0,l+1}, T_{1,l+1})$, $T_{0,l+1} > T_{1,l}$, when it moves towards its equilibrium, as the time behavior of a finite queueing system $GI \mid M \mid 1 \mid n_{l+1}$, which has a Poisson service distribution of parameter $\psi_{l,l+1}(n_l)$, $n_l = 1, \ldots, n_{l+1}$, and a general input distribution $B_{l+1}(x)$ equal to the service time distribution of resource R_{l+1}. This simplifies the analysis of $\mathscr{M}_{l+1}(n_{l+1})$ to obtain the stationary distribution $\{\pi_{l+1}(n_l \mid n_{l+1})\}$. The same procedure must of course be applied successively to all levels, starting from the lowest one. Only at this lowest level will the analysis be a bit more delicate since both R_0 and R_1 arbitrary service time distributions must be considered in $\mathscr{M}_1(n_1)$. This level-by-level procedure to analyze queueing networks with arbitrary service time distributions is all the more interesting due to the fact that no exact model yet exists to analyze such networks.

The conclusion that the output process of an aggregate $\mathscr{M}_l(n_l)$ is approximately a Poisson process of density $\psi_{l,k}(n_l)$, $k > l$, is not entirely dissociated from the superposition property evoked at the beginning of this section. If, for example, we consider expression (5.7), we see that, since the coefficients $A_l(j,n_l)$ are independent of time [cf. Eq. (5.8)], the density $\psi_{l,k}(n_l)$ may be regarded as the sum of the densities of $l+1$ renewal processes, each of density $A_l(j,n_l)$, $j = 0,\ldots,l$. Each such renewal process is associated with a component resource R_j of the aggregate; and these processes can be considered mutually independent because the aggregate is in equilibrium. The output process of an aggregate in equilibrium can in this way be viewed as being equivalent to the superposition of several independent renewal processes.

6.2.4 Fundamentally, the above properties follow from the fact that near-complete decomposability defines a scheme of decomposition both in time and in space; more precisely, Simon and Ando's properties of aggregate short-term equilibria justify the Poissonian assumption of time independency for certain particular behaviors of a system localized both in time and in space. The exploration of the near-complete decomposable character of a system may therefore lead to the validation in a model of the use of Poisson distributions, so widely used in other respects without other justification than the sheer necessity of overcoming analytic difficulties.

6.3 Further Generalizations

All the chapters that follow this one are entirely dedicated to the use of nearly completely decomposable models in the analysis of computer system performances. Before embarking upon this task, it is worthwhile to take a general standpoint and have at least a cursory look at some basic properties shared by all models built on the principles of aggregation and decomposability.

6.3.1 ***Heterogeneous Aggregation*** In the hierarchical model of aggregate resources developed in the preceding chapter, each aggregate $\mathcal{M}_l(n_l)$, at any level, has exactly the same structure and is therefore analyzed by the same single-server queueing model. This identity of aggregate structures is by no means a requirement of aggregation. Since the analysis of the short-term equilibrium may be carried out separately for each aggregate, it is possible, in principle, to amalgamate different models of aggregates within the framework of a single global nearly decomposable model. This possibility opens the way for the integration of different disciplines of analysis, e.g., queueing theory, simulation, or deterministic models, each aggregate being separately analyzed by the most suitable approach. Each aggregate model is, of course, constrained to yield output data tailored to the global model of interaggregate interactions. Besides, evaluation of the approximation will remain possible in these heterogeneous models provided that there are consistent means of assessing at least the minimum indecomposability of an aggregate and the maximum degree of coupling.

Aggregates of distinct structure will be encountered in the model of computer program behavior that are considered in Chapter VIII. Heterogeneous aggregation is probably the broadest and most promising field of application for nearly completely decomposable models.

6.3.2 ***Transient Behavior*** The principle of near-complete decomposability relates a partitioning of the state space of a model to a partitioning of its time behavior. This time–space relation has the useful effect of providing some information on the model transient behavior. In a multilevel nearly completely decomposable model, aggregates of successive levels reach their internal equilibrium states at successive time instants $T_{1,l}$, $l = 1, \dots, L$ (cf. Sections 1.3 and 1.5). For each such time instant the analysis yields the relative equilibrium values of the aggregative variables in terms of which the system is described at that level of aggregation.

This time succession of local equilibrium descriptions is in fact a discrete approximation, in time as well as in space, to the transient time behavior of the system. Such an approximation may be of great help if one considers the difficulty in obtaining the transient solutions of most mathematical models, even the simplest ones like, for example, the $M|M|1$ single-server queueing system (see, e.g., Cox and Smith, 1961, p. 61 ff). For no other reason than this difficulty, and again by sheer necessity of obtaining manageable solutions, many studies are inadequately restricted to a global steady-state analysis; naturally enough, these studies are accused of being unrealistic when there is no guarantee that the system, in practice, has ever time to come close to a stationary state. By demonstrating that many systems reach their global equilibrium by evolving through successive partial short-term equilibria, the theory of near-complete decomposability gives a clue as to which parts of the system the steady-state analysis should be applied and how steady-state results should be interpreted.

6.3.3 ***Balanced and Optimal Networks*** When one tries to assess the range of applicability of nearly completely decomposable models, the first question that comes to mind is whether the conditions for near-complete decomposability are or are not compatible with the conditions that are required for a system to be balanced or optimal with respect to some performance criterion. Without giving a definitive answer, we can at least make the following comments in the context of queueing networks.

By a *balanced* network we mean a network in which the frequency of access to a resource is proportional to the service rate of this resource, or more precisely, a network in which the following transfer probability and service rate ratios among resources are approximately equal:

$$\forall i, j, k: \quad p_{ik}/p_{ij} \approx \mu_k/\mu_j . \tag{6.16}$$

Balanced systems have the property that the long-term probability of a resource being idle or busy is approximately the same for all resources, regardless of their possibly quite different mean service rates and service time distributions. Indeed, take X_l, $l = 0, \ldots, L$, as the steady-state probability of at least one customer waiting or being served at resource R_l; in the basic model $\mathbf{Q}(N, L)$, these probabilities obey the following set of balance equations:

$$\sum_{l=0}^{L} X_l \mu_l p_{lk} = X_k \mu_k, \qquad k = 0, \ldots, L, \tag{6.17}$$

which express the condition that in the steady state the flow of customers in and out of each resource R_k must be equal. The system (6.17) defines the probabilities X_l only up to a multiplicative constant obtained by normalizing the sum to one.

Now, *the equality of the ratios* (6.16) *is a sufficient condition for all the probabilities* X_l *to be equal*. Suppose indeed that these ratios are equal; then from (6.17) we deduce

$$\sum_{l=0}^{L} X_l p_{lk} \mu_l / \mu_k = X_k, \qquad k = 0, \ldots, L,$$

or, since we have $\mu_l/\mu_k = p_{il}/p_{ik} = p_{ll}/p_{lk}$,

$$\sum_{l=0}^{L} X_l p_{ll} = X_k \qquad \text{for all} \quad k.$$

Note also that from $\sum_{k=0}^{L} p_{lk} = 1$, it follows that

$$\sum_{k=0}^{L} p_{ll} \mu_k / \mu_l = 1.$$

Thus,

$$p_{ll} = \mu_l \Big/ \sum_{k=0}^{L} \mu_k \qquad \text{or} \qquad \sum_{l=0}^{L} p_{ll} = 1,$$

i.e., the transfer probability matrix of a balanced network has a trace equal to unity.

Let us now see whether the ratios (6.16) conflict with or support the conditions of Theorem 4.1 and Section 4.3 for the near-complete decomposability of the model $\mathbf{Q}(N, L)$. Suppose the resources are ordered by decreasing service rates,

$$\mu_0 \geqslant \mu_1 \geqslant \cdots \geqslant \mu_L, \tag{6.18}$$

and remember (assuming $N \geqslant L+1$ for reasons of simplicity) that each level of aggregation l has a maximum degree of coupling ε_l, $l = 1, \ldots, L-1$, at most equal to ω_l, which is given [see (4.10)] by

$$\omega_l = \sum_{k=l+1}^{L} \mu_k \sum_{m=0}^{l} p_{km} + \sum_{k=0}^{l} \mu_k \sum_{m=l+1}^{L} p_{km},$$

and that the aggregate minimum indecomposability is proportional to $A_l + B_l$ with

$$A_l = \min_{0 \leqslant k \leqslant l-1} (\mu_k p_{kl}) \qquad \text{and} \qquad B_l = \mu_l \sum_{k=0}^{l-1} p_{lk}.$$

Because of the ordering (6.18), the ratios (6.16) have the effect, for each level l, of increasing A_l and B_l with respect to ω_l second summation, while keeping ω_l moderate by weighing the largest service rates μ_k, $k = 0, \ldots, l$, with the smallest probabilities p_{km}, $m = l+1, \ldots, L$, under the constraint that $\sum_{m=0}^{L} p_{km} = 1$. In this sense the conditions for near-complete decomposability are compatible with those required for a system to be balanced. More precisely, to achieve the same degree of near-complete decomposability, the service rate inequalities (6.18) must be less sharp in a balanced system than in one that is not balanced.

A balanced system is not necessarily an *optimal* system. On the contrary, for the central-server network model—its definition is given in Section 4.5—Buzen (1971a) proved that optimal system performance may only be achieved by soliciting the faster resources proportionately more than the slower ones. System performance can be defined in this system as the probability X_0 that the central server is busy. If there is a set of *other* resources R_i, $i \in \mathcal{R}$, that are functionally equivalent, the optimization problem consists in maximizing X_0 by adjusting the relative values of the probabilities p_{0i}, $i \in \mathcal{R}$, of a request from the central server to each of these functionally equivalent resources, under the condition that the total expected number of such requests per customer, i.e., the sum $\sum_{i \in \mathcal{R}} p_{0i}$, remains constant. Buzen showed that, for all pairs of resources (R_i, R_j) that belong to $\mathcal{R}$, the following conditions must hold at the operating point of optimal performance:

$$\begin{aligned} p_{0i}/p_{0j} &= \mu_i/\mu_j, \qquad \text{if} \quad \mu_i = \mu_j, \\ &> \mu_i/\mu_j, \qquad \text{if} \quad \mu_i > \mu_j. \end{aligned} \tag{6.19}$$

That is, to optimize the system throughput, resources with same service rate should receive the same share of requests, but faster resources should, with respect to their service rate, receive a larger proportion of requests than slower resources. This condition may be expected to remain valid for the more general network model $\mathbf{Q}(N, L)$. Consider a set $\mathcal{R}$ of functionally equivalent resources $R_k, R_j, \ldots, k, j \in \mathcal{R}$ and a resource R_i, $i \notin \mathcal{R}$. If we further assume that for all $k, j \in \mathcal{R}$, $p_{ki} \approx p_{ji}$, the optimal performance for R_i, i.e., maximum X_i in function of the p_{ik}, $k \in \mathcal{R}$, subject to the constraint that $\sum_{k \in \mathcal{R}} p_{ik}$ remains constant, would be obtained under the conditions

$$\begin{aligned} p_{ik}/p_{ij} &= \mu_k/\mu_j, \qquad \text{if} \quad \mu_i = \mu_j, \\ &> \mu_k/\mu_j, \qquad \text{if} \quad \mu_i > \mu_j. \end{aligned} \tag{6.20}$$

These conditions have the effect of increasing A_l and B_l with respect to ε_l even more than the ratios (6.16).

Optimization of resource utilization and, to a lesser extent, load balancing appear therefore as factors that promote near-complete decomposability in queueing networks.

6.4 Computational Efficiency

We have demonstrated in Sections 1.4 and 2.1 how a process of variable aggregation can reduce the resolution of an $n \times n$ nearly completely decomposable system of linear equations to the resolution of N smaller independent $n(I) \times n(I)$ systems. The total number of elementary arithmetic operations required to solve an $n \times n$ system is proportional to n^3 when classical and rigorous resolution techniques like Gaussian elimination and its derived methods (triangular factorization, Cholesky reduction, etc.) or the Gauss–Jordan elimination are used (see, e.g., Durieu, 1974). By aggregation, the number of operations required drops down to a magnitude of the order of $\sum_{I=1}^{N} n(I)^3 + N^3$, where $\sum_{I=1}^{N} n(I) = n$.

In queueing-network analysis, the gain is somewhat mitigated, at least in the case of the simple model $\mathbf{Q}(N, L)$. For this model Buzen (1973) proposed an efficient method to obtain the marginal probabilities $s_l(i_l)$ of i_l customers being in queue or in service at resource R_l:

$$s_l(i_l) = \frac{(X_l)^{i_l}}{G(N)} [G(N - i_l) - X_l G(N - i_l - 1)], \tag{6.21}$$

where the X_l are solutions of the system (6.17) and where $G(n)$ ($= 0$ for $n < 0$) is a normalizing constant defined as

$$G(n) = \sum_{\sum_{l=0}^{l=L} i_l = n} \prod_{l=0}^{l=L} (X_l)^{i_l}. \tag{6.22}$$

Buzen's iterative algorithm computes the entire set of values $G(1), \dots, G(N)$ using a total of $N \times (L+1)$ multiplications and $N \times (L+1)$ additions, these totals being multiplied by $N+1$ when service rates, as described in Section 6.1, are state dependent. One can compare these totals with the number of similar operations required by the hierarchial model of aggregate resources of Chapter V to compute the aggregative functions $\psi_{l,k}(n_l)$ defined by recurrence relation (5.5) or (6.5), and to compute the conditional distributions $\{\pi_l(n_{l-1} \mid n_l)\}$, $n_{l-1} = 0, \dots, n_l$, defined by (5.4) or (6.6). For each level l of aggregation, $L - l$ sets of N functions $\psi_{l,k}(n_l)$, $k = l+1, \dots, L$, $n_l = 1, \dots, N$, must be computed, each function requiring approximately n_l multiplications and n_l additions. This amounts to a total of approximately

$$N(N+1)/2 \times (L-1)L/2$$

additions plus as many multiplications. On the other hand, each of the N conditional distributions $\{\pi_l(n_{l-1} \mid n_l)\}$, $n_{l-1} = 0, \ldots, n_l$, $n_l = 1, \ldots, N$, needed at each level l requires $4n_l$ multiplications since each probability $\pi_l(n_{l-1} \mid n_l)$, $n_{l-1} = 1, \ldots, n_l$, can be obtained from $\pi_l(n_{l-1} - 1 \mid n_l)$ by the following recurrence formula derived from (5.4):

$$\pi_l(n_{l-1} \mid n_l)/\pi_l(n_{l-1} - 1 \mid n_l) = \mu_l(n_l - n_{l-1} + 1) \sum_{k=0}^{l-1} p_{lk} \Big/ \psi_{l-1,l}(n_{l-1}),$$

where $\pi_l(0 \mid n_l)$ is obtained simply by normalizing the sum to unity. Thus, the $N \times L$ conditional distributions for $n_l = 1, \ldots, N$, and for each level l, $l = 1, \ldots, L$, totalize $2N(N+1)L$ multiplications. Hence, altogether a total of additions and multiplications proportional to $(L^2 \times N^2)$ is necessary to calculate the aggregative variables $\psi_{lk}(n_l)$ and the conditional probabilities $\pi_l(n_{l-1} \mid n_l)$. And yet, in order to obtain the marginal probabilities $s_l(i_l)$ of the lower levels, we need to compute all those of the upper levels to be able to use the recurrent equations (5.14), while by (6.21), these marginal probabilities can be obtained separately from one another.

Compared in this way, the computational efficiency achieved by aggregation in the analysis of the model $\mathbf{Q}(N, L)$ appears rather poor. This comparison calls, however, for the following remarks. First, Eq. (6.22) requires the knowledge of the X_l, and thus the resolution of the system (6.17), the computation of which is proportional to $(L+1)^3$. If one remembers from Chapter IV that the smaller N, the better $\mathbf{Q}(N, L)$ near-complete decomposability is, it turns out that from the strict viewpoint of trading accuracy for computing time, the model of Chapter V is all the more competitive for L being large and N small. Second, and this is probably more important, it should be observed that if $\mathbf{Q}(N, L)$ is a nearly completely decomposable matrix, so is the matrix of system (6.17). Aggregation can thus be used to solve this system and obtain the X_l values, Buzen's algorithm being used to compute the set of functions $G(n)$ and to obtain by (6.2) the marginal congestion probabilities $s_l(i_l)$. This is in fact the approach that should be considered as far as network models as simple as $\mathbf{Q}(N, L)$ are concerned and when computational efficiency is the more critical issue. The hierarchical model of aggregate resources of Chapter IV has been considered essentially to unearth the basic relations that exist among levels of aggregation and among aggregate resources in networks of stochastic service systems. This hierarchical model finds its *raison d'être* in systems that, along the lines suggested in the earlier sections of this chapter, are more complex than $\mathbf{Q}(N, L)$ but where the same basic interlevel relations can be expected to hold. Examples of such systems will be encountered in the various applications that are considered in the remaining chapters.

CHAPTER VII

Memory Hierarchies

This chapter begins the third and final part of this book, which deals with the problem of evaluating computer system performances. In this chapter, we begin by analyzing the memory function of these systems; this function, as already stated in the Introduction, is a basic factor of hierarchical organization, and thus of near-complete decomposability in computer systems.

Most computer applications require memory systems with fast access and large capacity, which cannot be produced today at reasonable cost by one technology only. In any given technology fast access limits the memory capacity, and fast technologies are considerably more expensive than slow ones. The memory of a computer is thereby organized as a hierarchy of memory levels, say $M_0, M_1, \ldots, M_l, \ldots, M_L$. These levels, possibly realized with different technologies, are of increasingly larger capacity, and correspondingly of slower access speed and lower cost per unit of storage. Early examples of multilevel storage systems were the cache memory of the IBM System/360, Models 85 and 195 (Liptay, 1968); these systems had already used at least three memory levels ($L = 2$).

The goal of a hierarchical storage structure is to offer a total capacity equal to the sum of each level capacity, and at the same time to achieve an average access speed near the speed of the faster levels $M_0, M_1, \ldots$, for an average cost per unit of storage close to the cost of the higher, larger, and slower levels. This goal can be approached by means of an appropriate control of the movement of information across the levels, which aims at dynamically maintaining in the faster levels the subset of information currently needed by the programs in execution. To evaluate a hierarchical memory system is essentially to measure the success with which this goal has been attained.

Our first task in this chapter will be to set up a model of memory hierarchy that is tractable enough but also sufficiently general; we shall then examine the conditions under which this model can enjoy the property of near-complete decomposability. These conditions will turn out to conform with those which must be satisfied to use memory hierarchies efficiently. Then the model of aggregate resources set up in Chapter V will be used to define and evaluate performance criteria for storage hierarchies in which concurrent programs are executed on a multiprogrammed basis. This will enable us to evaluate the optimal number of multiprogrammed program executions (computations), which minimizes the average memory hierarchy access time, and also to evaluate the influence upon this average access time of parameters such as the capacity or the access time of a given individual memory level.

Many concepts and definitions used in the remainder of this work are introduced in this chapter, which in a way sets the stage for the following ones.

7.1 Basic Assumptions

There are many possible different implementations of a multilevel storage system. A model that would take all possible variants into account would be too impractical to obtain analytic results. We need therefore to impose some restrictions on the model we shall study. In this section, we take the utmost care to validate these restrictions.

7.1.1 Our model consists of $L+1$ memory levels $M_0, M_1, ..., M_l, ..., M_L$; we take the convention that slower levels correspond to higher subscripts. For $m < l$, M_m tends to be smaller, faster, and more expensive per unit of storage than M_l. We assume that level M_0 only has an access speed fast enough to match the processing speed of the central processor. Direct access of the processor to the hierarchy is thus restricted to M_0; information located at upper levels must be transferred to M_0 before it can be processed.

In certain multilevel storage systems, the processor has direct access to more than one fast level. Kaneko (1974), however, gives a procedure that permits the analysis of these systems to be reduced to the analysis of an equivalent hierarchy with only one level accessible to the processor; this single level has a capacity equal to the sum of the capacities of the actual directly addressable levels. Our assumption of a single directly addressable level is therefore not too constraining.

7.1.2 We assume that there exist direct data paths to move information down the hierarchy directly from each level M_l, $l > 0$, to M_0. Such data paths are needed to bring in the directly addressable levels the information requested by the processor within the shortest delays. We assume that information never transits via intermediate levels when moving down. We do not make any particular assumption on the paths information has to follow to move up in the hierarchy, but, as a consequence of the direct data paths mentioned above, we assume that this upstream information flow is independent of the downstream flow.

7.1.3 The time required to retrieve an element of information in a not directly addressable level is in practice much longer than the time needed to transfer this element to M_0. It becomes efficient then to transfer at once, not only the element, but also the context of information to which this element belongs, on the grounds that this context is likely to be needed by the program in the near future. Instead of being transferred element by element, information is thus transferred by blocks, called *pages*. Each memory level is divided into a number of *page frames*, each page frame being able to contain any one page. The page size is a critical parameter for a memory hierarchy. Too large a page may waste M_0 space with unrequested information, too small a page may not bring enough context and unduly increase the movement of pages across the hierarchy. In principle, the page size need not be the same for all levels. We shall, however, assume that it is; simple arguments (Knuth, 1968, p. 445 ff; Gelenbe, 1971) show that a single fixed page size not only simplifies the administration of page directories, but also minimizes the percentage of unused space.

We shall use c_l, $l = 0, \dots, L$, to denote the capacity of level M_l, i.e., its number of page frames.

7.1.4 A page located in an upper level M_l, $l > 0$, at the time it is referenced by the processor must be transferred to M_0 before it can be accessed. A replacement problem arises if M_0 is already full; a page must be evacuated from M_0 to make space for the new one. The page to be evacuated should ideally be of less immediate necessity to the program in execution than the other pages, and must logically be evacuated to an upper level of larger capacity, and cheaper and slower access. Note that the same replacement problem may arise at this upper level, which may also be full. Thus, for each memory level a *replacement policy* must be defined to select the "victim" that must be evacuated to an upper level whenever the level is full and a new page must come in. This policy may be different for each level. It is essentially through these replacement policies that the movement of pages across the levels can be controlled; a good policy should ideally anticipate the future program behavior and minimize the percentage of accesses to the higher, slower levels. In practice this is impossible. Various *replacement algorithms* have then been proposed to prognosticate the future behavior of programs by making predictions that rely on their immediate past behavior. These algorithms will be discussed in greater detail in the next chapter; let us simply mention here the more important ones: FIFO (selects the page that has remained for the longest time in the memory level), LRU (selects the page that has been the least recently referenced), LFU (selects the page that has been the least frequently referenced), WORKING SET (selects any page not referenced for a given number of past references). A good appraisal of these different strategies is given by Hoare and McKeag (1972).

7.1.5 An evaluation of the performances of a given memory hierarchy also requires a definition of the sequential order in which a program execution refers to its data, i.e., to its pages. The *page reference string* (Denning, 1970) is a model that provides this space–time relationship. In this chapter we shall use the term *computation* to refer to the dynamic process of a program in execution; the page reference string that identifies a computation ρ is simply a sequence of page references

$$\rho = r_1^{\rho}, r_2^{\rho}, \ldots, r_i^{\rho}, \ldots, r_{d(\rho)}^{\rho},$$

where it is understood that, if $r_i^{\rho} = y$, the computation ρ references its page y at the ith reference; y belongs to the computation page set and $d(\rho)$ is the length of the computation ρ, i.e., its total number of references.

We assume also that a computation is a strictly sequential process, i.e., that a reference cannot be made before the preceding one has been completed. Using $\tau(r_i^{\rho})$ to denote the time instant at which computation ρ makes its ith reference, we therefore assume that all page transfers initiated by a reference at a given instant $\tau(r_i^{\rho})$ are always completed before the time instant $\tau(r_{i+1}^{\rho})$ of the next reference of the *same* computation. As a consequence of this assumption, the page positions at instants $\tau(r_i^{\rho})$, $i = 1, 2, \ldots$, are independent of the actual time durations taken by the transfers between levels.

7.2 Access Probabilities

Let us now define f_l^{ρ} as the number of times a given page reference string ρ of length $d(\rho)$ makes a reference to a page located in level M_l, $l = 0, \ldots, L$, at the time it is referenced:

$$\sum_{l=0}^{L} f_l^{\rho} = d(\rho).$$

The *relative access frequency* to M_l of a given reference string ρ is then given by

$$F_l^{\rho} = d(\rho)^{-1} f_l^{\rho}, \qquad l = 0, \ldots, L,$$

with

$$\sum_{l=0}^{L} F_l^{\rho} = 1.$$

With F_l^ρ we can estimate the *access probability*, say p_l, of a page being located in M_l at the time it is referenced. Letting n be an arbitrarily large number of distinct reference strings ρ, $\rho = 1, \ldots, n$, each with access frequency F_l^ρ, we have

$$p_l = \lim_{n\to\infty} \left[n^{-1} \sum_{\rho=1}^{n} F_l^\rho \right], \qquad l = 0, \ldots, L, \tag{7.1}$$

with

$$\sum_{l=0}^{L} p_l = 1.$$

Alternatively, access probabilities could also be defined over a reference string of infinite length

$$p_l = \lim_{d(\rho)\to\infty} (F_l^\rho). \tag{7.2}$$

These access probabilities are a function of a large number of complex parameters: the reference string pattern, the replacement policies, the memory level capacities, the data paths between levels, etc. We must, however, remark that as a consequence of our basic assumptions, *the probability* p_l, $0 \leqslant l \leqslant L$, *is strictly a function of the reference string pattern and of the characteristics of the levels below and up to* M_l, *and does not depend on the characteristics of the levels above* M_l. Indeed, we know from the assumptions discussed in the previous section that (a) a page is brought into M_0 on demand only at the time it is referenced; (b) the downstream flow of pages does not impede the upstream flow; (c) a page pushed out from a given level is always evacuated to an upper level. Consequently, a referenced page found in a given level M_l comes necessarily from M_0 where it was at the time of its last reference, and has been pushed up successively through intermediate levels between M_0 and M_l by new pages being referenced. Hence the probability of a referenced page being found in a given level M_l is independent of the structure (data paths, capacity, replacement policy) of the levels above M_l.

The set of access probabilities $\{p_l\}$, $l = 0, \ldots, L$, is a definition of the stochastic behavior of a computation in a given memory hierarchy $M_0, \ldots, M_L$. We postpone to Section 7.9 the problem of their evaluation, and we shall see in the following sections how performance criteria for memory hierarchies can be defined in terms of these access probabilities. The main parameters we shall be interested in in these sections are the memory level capacities $c_0, c_1, \ldots, c_L$. For this reason, and because p_l is independent from the levels above M_l, it will be convenient henceforth to use $p_l(c_0, \ldots, c_l)$ to denote the access probability to level M_l.

7.3 Single-Process Storage Hierarchy

To evaluate the performances of a storage hierarchy the durations of the page transfers between levels must evidently be taken into account. We shall use $1/\mu_l$ to denote the mean time interval needed to retrieve and access a page in M_l, $l \neq 0$, and to transfer this page from M_l to M_0. The major component of $1/\mu_l$ is the time needed to retrieve and access the page. This is a characteristic of each memory level. We have assumed in Section 7.1.1 that levels are ordered in the hierarchy according to this characteristic in such a way that

$$\mu_1^{-1} \ll \mu_2^{-1} \ll \cdots \ll \mu_L^{-1}. \tag{7.3}$$

Similarly, we will use $1/\mu_0$ to denote the mean time needed by a computation to complete a reference to an element of information of a page located in M_0; we have evidently $\mu_0^{-1} \ll \mu_1^{-1}$.

We first assume in this section that one computation at a time is executed in the hierarchy. Since a computation is a strictly sequential process (our assumption 7.1.5), we may suppose that the time instant $\tau(r_{i+1}^\rho)$ at which this computation ρ initiates its $(i+1)$th reference is also the time instant at which the ith reference is completed. The mean time interval between two successive references of that computation is in this case

$$E\{\tau(r_{i+1}^\rho)-\tau(r_i^\rho)\} = \begin{cases} \mu_0^{-1}, & \text{if } \; r_i^\rho \in M_0 \qquad\qquad \text{at time } \; \tau(r_i^\rho), \\ \mu_0^{-1}+\mu_l^{-1}, & \text{if } \; r_i^\rho \in M_l, \quad l \neq 0, \quad \text{at time } \; \tau(r_i^\rho). \end{cases} \tag{7.4}$$

An *average hierarchy access time* $\overline{T}$ may then be defined as

$$\begin{aligned} \overline{T} &= p_0(c_0)\,\mu_0^{-1} + \sum_{l=1}^{L} p_l(c_0,\ldots,c_l)(\mu_0^{-1}+\mu_l^{-1}) \\ &= \mu_0^{-1} + \sum_{l=1}^{L} p_l(c_0,\ldots,c_l)\,\mu_l^{-1}, \end{aligned} \tag{7.5}$$

and an *average fraction of time* η *lost in transferring pages between any level* M_l, $l \neq 0$, *and* M_0, as

$$\eta = (\overline{T}-\mu_0^{-1})/\overline{T}. \tag{7.6}$$

7.4 Multiprocess Storage Hierarchy

Suppose now that transferring pages from M_l, $l \neq 0$, to M_0 and servicing references in M_0 may be performed simultaneously. Suppose also that memory space at the various levels is equally, statically, and permanently shared among N independent computations ρ, $\rho = 1, \ldots, N$, executed on a multiprogramming basis. This means that the execution of the N corresponding reference strings is interleaved in the following way: iff $r_i^\rho \in M_l$, $l \neq 0$, at instant $\tau(r_i^\rho)$ and iff there exists a computation ρ' of which $r_j^{\rho'}$ is the last reference completed and for which $r_{j+1}^{\rho'} \in M_0$ at some time $t > \tau(r_i^\rho)$ before the transfer of the page r_i^ρ is completed, then $r_{j+1}^{\rho'}$ is serviced at time t.

This scheme provides for a better utilization of M_0 if the hardware is such that references may be serviced in M_0 while pages are being transferred from the upper levels to M_0. As a consequence of this, however, the performance measures defined by (7.5) and (7.6) are no longer valid: they require indeed that only one reference to any one level M_l, $l = 0, \ldots, L$, be made and completed at a time, whereas under multiprogramming the time intervals $[\tau(r_i^\rho), \tau(r_{i+1}^\rho)]$ of distinct computations ρ may overlap each other.

In a multiprogramming storage hierarchy, the average time during which M_0 is not being referenced within a time interval $[\tau(r_i^\rho), \tau(r_{i+1}^\rho)]$ may be less than μ_l^{-1} when $r_i^\rho \in M_l$, $l \neq 0$, at time $\tau(r_i^\rho)$. On the other hand, as one page only may be transferred from a level M_l, $l \neq 0$, to M_0 at a time, requests for page transfers may eventually queue up at each level, and time intervals $[\tau(r_i^\rho), \tau(r_{i+1}^\rho)]$ will be prolonged by these queueing times.

What will be, therefore, the effective gain obtained by multiprogramming? Which conditions should the access probability distribution $\{p_l(c_0, \ldots, c_l)\}_{l=0}^L$ obey in order to have an optimal average hierarchy access time? We know already from Section 6.3.3 that an optimal hierarchy should not be a balanced one and that faster levels should be solicited proportionately more frequently than slower ones. But how much more frequently? To answer such questions we need performance criteria, comparable to (7.5) and (7.6), but valid under multiprogramming.

7.5 Near-Complete Decomposability

The reader will have already understood from inequalities (7.3) how a memory hierarchy can be viewed as a nearly completely decomposable system. Let us just make this point more precise. We can identify each memory level M_l with a resource R_l with exponentially distributed service times of mean μ_l^{-1}, and each of the N multiprogramed computations with a sequential process stochastically defined by the same probability set $\{p_l(c_0/N, \ldots, c_l/N)\}_{l=0}^L$. Let i_l, $l = 1, \ldots, L$, $i_l = 0, \ldots, N$, be the number of computations that at some time t are waiting for a page to be transferred from M_l, $l \neq 0$, to M_0, and let

$$i_0 = N - \sum_{l=1}^{L} i_l.$$

Defined in this way, this model of a N-process storage hierarchy is a closed network of queues, the state of which is defined by the $(L+1)$-tuple $(i_0, \ldots, i_L)$, and the transition stochastic matrix of which is $\mathbf{Q}(N, L)$ (cf. Chapter IV).

More precisely, the network is the central server queueing network introduced in Section 4.5, since the transfer probabilities p_{lk} between resources are here given by

$$\begin{aligned} \forall k, l \neq 0: \quad p_{0k} &= p_k(c_0/N, \ldots, c_k/N), \\ p_{l0} &= 1, \qquad p_{lk} = 0, \qquad p_{ll} = 0, \\ p_{00} &= 1 - \sum_{k=1}^{L} p_{0k}. \end{aligned} \tag{7.7}$$

Sufficient conditions for the $(L-1)$-level near-complete decomposability of this system into subsystems $\mathbf{Q}(n,l)$ were given in Section (4.5). We must have, for $l = 1, \ldots, L-1$,

$$\omega_l \leqslant \tfrac{1}{2}(A_l + B_l) - (A_l B_l)^{1/2} \cos[\pi/(N+1)], \tag{7.8}$$

where ω_l, A_l, and B_l are

$$\omega_l = \max_{\substack{i_0, \ldots, i_L \\ \sum i_l = N}} \left[\sum_{k=l+1}^{L} \kappa(i_k)\mu_k + \kappa(i_0)\mu_0 \sum_{m=l+1}^{L} p_m(c_0/N, \ldots, c_m/N) \right],$$

$$A_l = \mu_0 p_l(c_0/N, \ldots, c_l/N), \qquad B_l = \mu_l.$$

Conditions (7.7) are thus expressed in terms of rates at which a page may be retrieved, accessed, and transferred from each level, and in terms of probabilities with which levels are referenced by an individual computation. When these conditions are satisfied, each memory level above M_0 corresponds to a level of aggregation.

One may expect inequalities (7.8) to hold in most storage hierarchies of present computer systems for various reasons. An essential reason is that the access times of the different types of storage devices differ by at least one order of magnitude from one level to the other. As an example, some typical values of average access times are given in Table 7.1.

If we assume in first approximation that $\mu_l \approx \mu_0 p_l(c_0, \ldots, c_l)$, $l = 1, \ldots, L$,

TABLE 7.1

AVERAGE MEMORY ACCESS TIMES

Storage device	μ_l^{-1}
CACHE	10^{-4} msec/reference
CORE	10^{-3} msec/reference
DRUM (head per track disk)	10–30 msec/page
DISK	100–1000 msec/page
MAGNETIC TAPE	1000–100,000 msec/page
MASS TAPE	50,000 msec/page

these values indicate that the rate of interaction *within* the aggregates of a same level will be for each level at least ten times higher than the rate of interaction *between* aggregates. From our discussion in Section 4.4, we know that this ratio is sufficient to guarantee that inequalities (7.8) hold for values of the degree N of multiprogramming up to between 5 and 10 (see Table 4.1). There is another important reason for inequalities (7.8) to hold. The lower bound for the average time needed to complete a single reference of a *given* computation corresponds to the case of no queueing of page transfer requests for this computation and is equal to

$$\mu_0^{-1} + \sum_{l=1}^{L} p_l(c_0/N, \ldots, c_l/N)\mu_l^{-1}. \tag{7.9}$$

To optimize the average total execution time of a *given* computation, it is mandatory to minimize (7.9). Both the user and the system storage management policies will aim at this insofar as they have control over the placing of information in the various levels, i.e., over the values of the access probabilities. But, as a result of inequalities (7.3), the only possible way to reduce (7.9) is to aim at the set of inequalities

$$p_0(c_0/N) \gg p_1(c_0/N, c_1/N) \gg \cdots \gg p_L(c_0/N, \ldots, c_L/N). \tag{7.10}$$

And these inequalities (7.10) have, for each level l, the effect of increasing A_l with respect to ε_l, and thus to support conditions (7.8).

Still another support for conditions (7.8) is the observation that computations enjoy what has been called the *locality property* (Belady, 1966; Denning, 1968a), that is, programs show a tendency to favor at each instant of their execution a subset of their information only, instead of scattering their references uniformly over their total set of information; they tend to use for relatively long periods of time rather small subsets of information only. This property had already been experimentally observed by Naur (1965), and also by Fine *et al.* (1966), Varian and Coffman (1967), and Madison and Batson (1975). This behavior is the result of program loops, iterations, and recursions; it is also the result of procedure and subroutine executions, of the gathering of data into segments to which content-related references are made in clusters. A consequence of this pattern of localized references is that every page transfer that is occasioned by a single reference to an upper level is likely to generate, on the average, a large number of references to M_0. Thus the probability $p_0(c_0/N)$ will tend to be comparatively larger than the other access probabilities, and this is again a factor of near-complete decomposability.

When the favored subsets can be detected (we shall refer to them as *localities*) it becomes possible to execute a program within a limited portion of M_0 and with a tolerable loss of efficiency (i.e., with a tolerable number of accesses to the higher and slower levels) by maintaining in M_0 only the locality currently used by a computation. This detection of localities is in fact the essential objective of the page replacement policies such as those mentioned in Section 7.1.4; they aim at maintaining the localities in execution in the lower levels of the hierarchy by choosing their "victim" outside these localities. This also has the effect of sharpening the inequalities (7.10) and thus of contributing to the near-complete decomposability of the hierarchy structure.

We can therefore conclude that conditions (7.8) are almost an inborn property of storage hierarchies and are likely to be satisfied in most practical cases. This will be corroborated by the numerical and detailed case studies treated in the following chapters.

7.6 Memory Level Aggregation

The model of a hierarchy of aggregate resources that was studied in Chapter V is applicable to an N-process storage hierarchy that verifies conditions (7.8, and performance criteria for this storage hierarchy that are comparable to (7.5) and (7.6) can be defined within this model.

The aggregate $\mathcal{M}_l(n_l)$, $n_l = 1, \ldots, N$, $l = 1, \ldots, L$, defined in Chapter V can be regarded as a storage hierarchy of levels $M_0, \ldots, M_l$ in which n_l computations, defined by the same set of access probabilities $\{p_k(c_0/N, \ldots, c_k/N)\}_{k=0}^{l}$, are either served in M_0 or waiting for a page transfer from levels $M_1, \ldots, M_l$. Capacities $c_0, \ldots, c_l$ are divided by N and not by n_l, because to start with we assume that a fixed and equal portion of memory space is permanently allocated at every level to each of the N multiprogrammed computations. We define $\sigma_l(n_l)$ as the mean number of references serviced from M_0 per time unit within this aggregate $\mathcal{M}_l(n_l)$; $\sigma_l(n_l)$ can be obtained by a recurrence relation similar to (5.11):

$$\sigma_l(n_l) = \sum_{n_{l-1}=1}^{n_l} \pi_l(n_{l-1} \mid n_l)\, \sigma_{l-1}(n_{l-1}), \qquad \text{for} \quad l = 1, \ldots, L, \tag{7.11}$$

$$\sigma_0(n_0) = \begin{cases} \mu_0 & \text{if} \quad n_0 \neq 0 \qquad \text{for} \quad l = 0, \\ 0 & \text{otherwise.} \end{cases}$$

The conditional probabilities $\pi_l(n_{l-1} \mid n_l)$, $n_{l-1} = 0, \ldots, n_l$, are yielded by relations (5.4), which because of (7.7) reduce to

$$\pi_l(n_{l-1} \mid n_l) = \mu_l^{n_{l-1}} \left[\prod_{k=1}^{n_{l-1}} \psi_{l-1,l}(k) \right]^{-1} \pi_l(0 \mid n_l), \tag{7.12}$$

where $\pi_l(0 \mid n_l)$ is obtained from the condition

$$\sum_{n_{l-1}=0}^{n_l} \pi_l(n_{l-1} \mid n_l) = 1$$

and where $\psi_{l-1,l}(k)$, defined as the rate at which aggregate $\mathcal{M}_{l-1}(k)$ references memory level M_l, is given by the recurrence relation (5.5), which with (7.7) reduces to

$$\psi_{l-1,l}(k) = \sum_{n_{l-2}=1}^{k} \pi_{l-1}(n_{l-2} \mid k)\, \psi_{l-2,l}(n_{l-2}), \tag{7.13}$$

with $\psi_{0,l}(n_0) = \mu_0 \times p_l(c_0/N, \ldots, c_l/N)$ for all $n_0 \neq 0$.

An average access time $\overline{T}_N$ for an N-process storage hierarchy can now be obtained as *the inverse of the average number of references serviced from* M_0 *per time unit within the uppermost level aggregate* $\mathcal{M}_l(N)$:

$$\overline{T}_N = \sigma_L^{-1}(N). \tag{7.14}$$

Moreover, the average fraction of time during which all N computations are waiting for a page transfer from some memory level M_l, $l \neq 0$, is given by

$$\eta_N = s_0(0), \tag{7.15}$$

where $s_0(i_0)$ is the long-term and unconditional probability of i_0 computations in service or in queue at level M_0 and is defined by expressions (5.14).

One can prove that η_N and $\overline{T}_N$ satisfy the same relation (7.6) that was satisfied by η and $\overline{T}$, namely,

$$\eta_N = (\overline{T}_N - \mu_0^{-1})/\overline{T}_N. \tag{7.16}$$

PROOF From (5.15) we have

$$\begin{aligned} s_0(0) &\equiv a_1(0) \\ &= 1 - \sum_{n_0=1}^{N} a_1(n_0). \\ &= 1 - \sum_{n_{L-1}=1}^{N} a_L(n_{L-1}) \sum_{n_{L-2}=1}^{N_{L-1}} \pi_{L-1}(n_{L-2} \mid n_{L-1}) \cdots \sum_{n_0=1}^{n_1} \pi_1(n_0 \mid n_1). \end{aligned}$$

By virtue of (7.11) we have

$$\sum_{n_0=1}^{n_1} \pi_1(n_0 \mid n_1) = \mu_0^{-1} \sigma_1(n_1),$$

and applying (7.11) successively for $l = 2, \ldots, L-1$, we obtain

$$a_1(0) = 1 - \mu_0^{-1} \sum_{n_{L-1}=1}^{N} a_L(n_{L-1})\, \sigma_{L-1}(n_{L-1}) = 1 - \mu_0^{-1} \sigma_L(N)$$

which completes the proof.

It is interesting to verify also that our aggregative model of an N-process storage hierarchy is compatible with our model of a single-process hierarchy, i.e., that for $N = 1$, $\overline{T}_N$ and η_N reduce to

$$\overline{T}_1 = \overline{T} \qquad \text{and} \qquad \eta_1 = \eta. \tag{7.17}$$

PROOF We obtain from (7.12) that, in the queueing system $\mathcal{M}_l(1)$, $l=1,\dots,L$, $\pi_l(1|1)$ can be expressed as

$$\pi_l(1|1) = \mu_l/[\mu_l+\psi_{l-1,l}(1)].$$

Similarly, from (7.13) and (7.11) for $N=1$ we obtain

$$\psi_{l-1,l}(1) = \prod_{k=l-1}^{1} \pi_k(1|1)\, p_l(c_0,\dots,c_l)\,\mu_0,$$

$$\sigma_{l-1}(1) = \prod_{k=l-1}^{1} \pi_k(1|1)\,\mu_0.$$

These equations yield

$$\pi_l(1|1) = \mu_l/[\mu_l+p_l(c_0,\dots,c_l)\,\sigma_{l-1}(1)].$$

Thus,

$$\begin{aligned}\sigma_l^{-1}(1) &= \pi_l^{-1}(1|1)\,\sigma_{l-1}^{-1}(1)\\ &= \sigma_{l-1}^{-1}(1) + p_l(c_0,\dots,c_l)\,\mu_l^{-1}.\end{aligned}$$

Applying this recurrence L times and remembering that $\sigma_0(1)^{-1} = \mu_0^{-1}$, one finds

$$\overline{T}_1 = \sigma_L^{-1}(1) = \sum_{l=1}^{L} p_l(c_0,\dots,c_l)\,\mu_l^{-1} + \mu_0^{-1} = \overline{T};$$

and thus, by (7.16), we have also $\eta_1 = \eta$.

Furthermore, let us remark that the *utilization factor* of the storage hierarchy channel that performs transfers from M_l to M_0, $l \neq 0$, can be obtained as $1-s_l(0)$, probability $s_l(0)$ being defined by (5.14). Likewise (5.17) defines the mean response time $\overline{W}_l$ of this component.

Finally, if the uppermost level M_L is the model of some infinite reservoir from which all computations originate at a rate given by (4.2) and to which they all return when completed, then W, the mean response time of the aggregate consisting of $M_0, M_1, \dots, M_{L-1}$, is defined by (5.18); it is the mean time required to service all references of a reference string, or the mean time to execute a computation.

7.7 Dynamic Space Sharing

The performance improvement achieved by multiprogramming N computations so as to take advantage of the simultaneity between transfers between memory levels and servicing references in M_0 can now be measured by the ratio

$$\overline{T}_1/\overline{T}_N \qquad \text{or} \qquad \eta_N/\eta_1. \tag{7.18}$$

In general, there will exist one optimal value N_{opt} that maximizes these performance ratios. This optimum results from two counteracting effects: on the one hand, the probability $1 - s_0(0)$ of having at least one computation not waiting for a page transfer from some upper level M_l, $l \neq 0$, to M_0 is inclined to increase with N; on the other hand, the average memory space c_l/N, $l = 0, 1, \ldots$, available to each computation at the lower levels shrinks as N increases, with the consequence that the probabilities $p_k(c_0/N, \ldots, c_k/N)$, $k > l$, of accessing the upper levels increase rapidly. The probabilistic model described above provides a means to evaluate this optimum, given (i) a family of programs defined by their reference strings from which access probabilities may be inferred by one of the techniques that will be discussed in Section 7.8, and (ii) a nearly completely decomposable memory hierarchy of an arbitrary depth, each memory level being characterized by its capacity and a distribution of the amount of time required to access, retrieve, and transfer a page from this level to the executive memory.

So far we have assumed that a fixed and equal portion c_l/N, $l = 0, \ldots, L$, of memory space was permanently allocated at every level to each of N computations in the hierarchy. This is, however, not the case in most hierarchical storage systems. Usually, in order to keep the access probabilities to the upper levels moderate, an upper limit, say J_{max}, is imposed upon the number of computations J allowed to share space in the lower and faster memory levels, say $M_0, \ldots, M_A$. Any of these computations is expropriated from the space it has accumulated in these levels as soon as it makes a reference to or above a certain level, say, M_B, $B > A$. Thus, if n_{B-1} designates the number of computations not waiting for a page transfer from levels $M_B, \ldots, M_L$, we have

$$J = \min(n_{B-1}, J_{max}),$$

and the average space allotted to any of these computations at any time and at any level M_l, $l \leqslant A$, is the integer part of c_l/J.

This space allocation policy preserves the near-complete decomposability property of the hierarchy since its net effect is to enhance the difference between access probabilities to the lower levels $M_0, \ldots, M_A$ and the upper levels $M_{A+1}, \ldots, M_L$. We must consider J_{max} distinct access probability distributions $D(J)$, conditioned on the value of n_{B-1}:

$$D(J): \begin{cases} p_l(c_0/\alpha_0, \ldots, c_l/\alpha_l), & l = 1, \ldots, L, \\ p_0 = 1 - \sum_{l=1}^{L} p_l(c_0/\alpha_0, \ldots, c_l/\alpha_l), \end{cases} \tag{7.19}$$

where $J = 1, \ldots, J_{max}$, and

$$\alpha_l = \begin{cases} J, & \text{for } l = 0, \ldots, A, \\ N, & \text{for } l = A + 1, \ldots, L. \end{cases}$$

There are three stages in the analysis of such a hierarchy:

(i) The bottom levels $M_0, \ldots, M_A$ are modeled by $J_{\max}$ aggregates $\mathcal{M}_A(J)$, $J = 1, \ldots, J_{\max}$, each with the appropriate access probability distribution $D(J)$. The analysis of these aggregates yields the interaction rates $\psi_{A,l}(J)$ of the bottom levels upon each level $l = A+1, \ldots, L$.

(ii) Next, one considers N aggregates $\mathcal{M}_{B-1}(n_{B-1})$, $n_{B-1} = 1, \ldots, N$. The interaction rate of the bottom levels upon any one intermediary level M_i, $i = A+1, \ldots, B-1$, in each of these aggregates is taken equal to $\psi_{A,i}(n_{B-1})$ if $n_{B-1} < J_{\max}$, or to $\psi_{A,i}(J_{\max})$ otherwise.

(iii) The analysis of each aggregate $\mathcal{M}_{B-1}(n_{B-1})$ yields the interaction rates $\psi_{B-1,l}(n_{B-1})$ upon the upper levels l, $l = B, \ldots, L$; using these interaction rates with (5.4), the upper levels may be modeled by a system $\mathcal{M}_L(N)$.

The same argument that leads to an optimal value for N is applicable to $J_{\max}$. The model outlined above may be used to estimate the value $J_{\max}^{\text{opt}}$ of an optimal *maximum degree of multiprogramming* in the lowest memory levels, which for a given N value maximizes the ratios (7.18). Such an optimal value $J_{\max}^{\text{opt}}$ will be calculated in Chapter IX for a computer system using the space allocation policy discussed above.

This policy of using state-dependent access probabilities (i.e., state-dependent transfer probabilities) to take into account dynamic space sharing among multiprogrammed computations is an advantage over previous comparable approaches such as the one followed recently by Gecsei and Lukes (1974).

7.8 Linear Storage Hierarchies

Performance criteria such as the hierarchy average access time or the processor utilization have been defined in terms of access probabilities. Thus, when the goal of the analysis is to find for a parameter an optimal value that maximizes these performance criteria, e.g., a best replacement strategy or an optimal combination $(c_0, \ldots, c_L)$ of memory capacities, the prior knowledge of all possible sets $\{p_l(c_0, \ldots, c_l)\}_{l=0}^{L}$ is required, one set for each possible value of the varying parameter. This may represent an enormous task, especially when an optimal combination $(c_0, \ldots, c_L)$ is sought, and when simulation and/or measurements are the only feasible methods. Access probabilities are indeed a much condensed representation of the behavior of the computations, a behavior that is itself a complex function of many parameters (data paths, replacement policies, capacities, reference string patterns). This task can, however, be considerably reduced when the storage hierarchy is *linear* and when the *same* replacement algorithm is used for all memory levels.

A linear hierarchy is a hierarchy in which a page replaced at any level M_l is always moved to the *next* upper level M_{l+1} (Mattson *et al.*, 1970). *In such a hierarchy, when the same arbitrary replacement algorithm is applied at all levels, the access probability* $p_l(c_0, \ldots, c_l)$ *depends only on* c_l *and on the total capacity* $c_0 + c_1 + \cdots + c_{l-1}$. To show this, let us suppose that a page is being referenced at a certain level M_l. From all lower levels $M_0, M_1, M_2, \ldots, M_{l-1}$ that are full a page must be replaced and moved to the level immediately above until a vacant page frame is found. This vacant frame will certainly be found, if not below, at least in M_l where space was left free by the page just referenced. At each full level $M_0, M_1, \ldots$ the replacement algorithm selects the page for replacement from the set consisting of the pages already present and of the page being pushed up from the level below. As the criterion for replacement is the same for all levels, the page with highest priority for replacement at some intermediate level M_k is thus also the page with highest priority for replacement in M_k and M_{k-1}; the same argument can be applied from level to level to all lower levels so that this page has also highest priority for replacement in all levels $M_k, \ldots, M_0$. This page can thus be viewed as being selected from a unique memory level of total capacity $c_0 + c_1 + \cdots + c_k$. Therefore, as far as the position of pages in the upper levels M_k, $M_{k+1}, \ldots, M_L$ is concerned, our linear hierarchy is equivalent to, i.e., has same behavior as, a reduced hierarchy of $L-k+1$ levels only: the lowest level, say M_k', of capacity $c_0 + \cdots + c_k$, and the levels $M_{k+1}, \ldots, M_L$. For the same set of reference strings, the access probabilities $p_l(c_0, \ldots, c_l)$, $l = k+1, \ldots, L$, will be the same in both hierarchies.

This equivalence simplifies the evaluation of the access probabilities. Let us define

$$C_k = c_0 + c_1 + \cdots + c_k, \qquad k = 0, \ldots, L,$$

and

$$F(A, C_k) = p_{k+1}(c_0, \ldots, c_{k+1}) + \cdots + p_L(c_0, \ldots, c_L), \qquad k = 0, \ldots, L-1.$$

$F(A, C_k)$ is, for a given replacement algorithm A, the probability that a page being referenced is located in one of the upper levels $M_{k+1}, \ldots, M_L$. It is also the probability that a page being referenced is not in the lowest level M_k' of capacity C_k of the equivalent reduced hierarchy. In other words, $F(A, C_k)$ is, for a given replacement algorithm A, *the probability of a page fault* in a hierarchy lowest level of capacity C_k. From such page fault probabilities, the access probabilities can be easily derived:

$$p_k(c_0, \ldots, c_k) = F(A, C_{k-1}) - F(A, C_k), \qquad k = 1, \ldots, L,$$

with

$$p_L(c_0, \ldots, c_L) = F(A, C_{L-1}), \qquad p_0(c_0) = 1 - \sum_{l=1}^{L} p(c_0, \ldots, c_l). \quad (7.20)$$

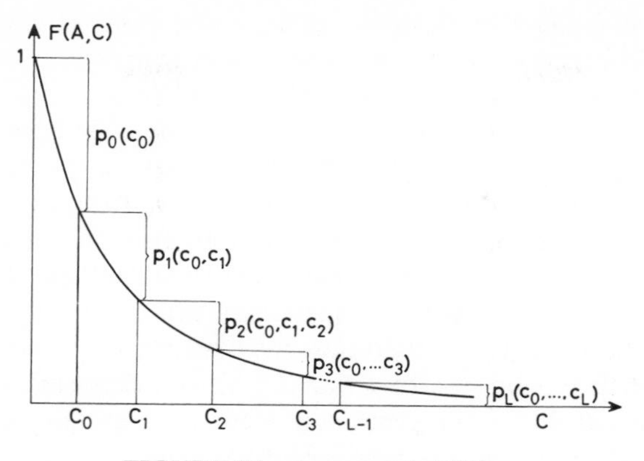

FIGURE 7.1 Access probabilities.

For a given algorithm A, $F(A, C_k)$ depends on C_k only; and thus, as shown by Eq. (7.20), $p_k(c_0, \ldots, c_k)$ depends on the total capacity C_{k-1} and on c_k only in a linear hierarchy with unique replacement algorithm. This property reduces the amount of calculation required to obtain the various sets $\{p_k(c_0, \ldots, c_k)\}_{k=0}^{L}$ that may be needed when some optimal combination $(c_0, \ldots, c_L)$ has to be found. It suffices to calculate a page fault probability $F(A, C)$ for all possible integer values that may be assigned to C_k, $k = 0, \ldots, L$; as indicated by Fig. 7.1, access probabilities $p_l(c_0, \ldots, c_l)$ can then be readily obtained from the function $F(A, C)$ for any relevant combination $(c_0, \ldots, c_l)$.

The remaining problem is evidently the determination of the page fault probability $F(A, C)$ in function of A and C. This is one of the topics dealt with in the next chapter, where we shall see that $F(A, C)$ can be derived by calculation for most replacement algorithms from a stochastic definition of program-paging behavior.

Of course, the above technique works provided that the probability $F(A, C)$ is a nonincreasing function of C; otherwise it yields negative probabilities [Eq. (7.20)]. But this condition will be verified in most practical cases, and certainly if the replacement algorithm belongs to the class introduced by Mattson *et al.* (1970) and called *stack algorithms*. The set of pages maintained by a stack algorithm in a memory of capacity C is always a subset of the set of pages it would maintain in a memory of capacity $C+1$; increasing the memory capacity can never in this case increase the probability of a page fault. Algorithms such as LRU, LFU, or WORKING SET, which use a replacement criterion independent of the memory capacity, are all stack algorithms. FIFO on the contrary is an example of a nonstack algorithm. For some pathological sequences of references, FIFO has the unexpected effect of increasing the number of page faults with C (Belady *et al.*, 1969); such sequences, however, remain rare in practice so that in general one can expect the probability $F(\text{FIFO}, C)$ also to be a nonincreasing function of C.

Equation (7.20) and the equivalence property discussed above were inspired and somewhat generalized in terms of probabilities and for all replacement algorithms by similar results already obtained by Mattson *et al.* (1970); they proposed a stack-processing technique that determines, in a single simulation pass of an address trace, the access frequencies to all levels of a linear storage hierarchy with unique replacement algorithm, provided that this algorithm is a stack algorithm. This last condition has been replaced here by the somewhat less restrictive condition that $F(A, C)$ be nonincreasing with C. When conditions are satisfied, stack-processing techniques are, of course, efficient techniques for the evaluation of access probabilities.

CHAPTER VIII

Near-Complete Decomposability in Program Behavior

8.1 Preliminaries

We have seen in the preceding chapter how the efficiency of multilevel storage systems could be evaluated in terms of access probabilities; and how, in the case of a linear memory hierarchy with unique page replacement policy, these access probabilities are simple functions of the page fault probability in an equivalent reduced two-level hierarchy.

We shall now focus our attention on the calculation of the page fault probability and of other similar program characteristics. These characteristics depend on memory parameters and page replacement policies; but they also depend strongly on the patterns followed by programs to reference their paged information.

The calculation of these characteristics requires therefore a realistic and tractable model of program paging behavior. Several models have been proposed—they will be briefly discussed in the following section—but curiously enough, most of them fail to give a true account of an important property of program paging behavior: the locality property (Section 7.5). They fail to reproduce the two very different types of behavior exhibited by a program when it is referencing within a locality and when it is changing locality.

In this chapter we investigate an approach by which this deficiency may be overcome, at least under certain assumptions; cases will remain for which our model can only suggest the lines along which further studies could proceed.

We have said already that factors that contribute to locality are program iterations and recursions; also the organization of data into segments to which content-related references are made in clusters. But, more fundamentally, program locality probably finds its origin in the design and structuring process itself of the program. Like many other complex organized systems, a program is usually conceived as an assembly of interrelated modules. Each module is responsible for some specific function and operates upon a restricted subset of data. If the function is complex enough, the module will itself be an assembly of interrelated submodules, until some lowest level of elementary constituents is attained. The idea behind such a decomposition is to give each module a maximum functional autonomy: submodules of a same module are closely dependent upon each other, but submodules of distinct modules are not. The likely effect of this organization, when it is in operation, is that the rate of interaction within a module will generally be higher than between modules. In other words, when a module is being executed, it is likely to operate within its restricted domain of data for a substantially long interval—measured in terms of elementary operations (instructions)—before communication with another module is required. We can thus expect such modules to correspond to the aggregate of a nearly completely decomposable structure, and the set of data accessible to a module (submodule) to correspond to a locality (sub-locality).[1] It was therefore natural to consider the assumption of near-complete decomposability as a promising starting point for a model of program locality.

Our approach will be illustrated by a numerical example. This example shows another important advantage of aggregation. Most models that have been presented in the literature require amounts of computer time and space that are beyond practical possibilities for normal program page set sizes. Aggregation keeps these requirements within feasible limits.

Much of the material presented in this chapter is drawn from Courtois and Vantilborgh (1976).

8.2 Existing Models of Program Paging Behavior

We have already used two representations of program paging behavior: the reference string (Section 7.1.5) and the set of access probabilities (Section 7.2). The reference string is an exact copy of the original paging behavior of a program but, as such, can be exploited for measurement or simulation only. The access probabilities are, on the other hand, already the result of a complex analysis of the behavior of programs in a given hierarchical storage system. The models we are considering in this chapter are at a level of abstraction that is between these two extremes. They aim at abstracting from the reference strings the essential properties of program behavior, and they consist of simplifying assumptions that make possible the evaluation of characteristics

[1] Incidentally, this discussion shows that structure and efficiency are two qualities of a program that are not necessarily mutually exclusive.

such as the page fault probability, the access probabilities, etc. In this section, we review the more important models of this type that have been proposed in the literature, and focus our attention on the way they take into account the locality phenomenon. A more comprehensive discussion of these existing models can be found in Courtois and Vantilborgh (1976).

8.2.1 ***The Independant Reference Model*** **(IRM)** This model has been much exploited because it has the best mathematical tractability. The behavior of a program is represented by an infinite reference string (Section 7.1.5): $r_1 \cdots r_t \cdots r_{t'} \cdots r_{t''} \cdots$; it is understood that if $r_t = i$, the program's tth reference is directed to page i of the program page set $\{1, \ldots, n\}$. The IRM model is then based on the following assumptions:

(i) For all t and $1 \leqslant i \leqslant n$,

$$\Pr\{r_t = i\} = b_i,$$

with $b_i > 0$ and $\sum_{i=1}^{n} b_i = 1$.

(ii) For all $t' \neq t''$ and $1 \leqslant i, j \leqslant n$, $\{r_{t'} = i\}$ and $\{r_{t''} = j\}$ are independent events.

Thus, each page is supposed to be referenced with a constant probability b_i, which remains independent of time and of the references addressed to other pages. These assumptions make it possible to obtain closed-form expressions for the page fault probability of most classical replacement algorithms (King, 1971; Aho *et al.*, 1971). Let us take the LRU (least recently used) replacement algorithm as an example; in this chapter we shall use $F(A, C, n)$ to denote the stationary probability of a page fault, i.e., the stationary probability that the page being referenced is not present in a main memory, given that this memory may contain at most C distinct pages, that A is the replacement algorithm, and n the number of pages of the program:

$$F(A, C, n) = \lim_{t \to \infty} \Pr\{r_t \notin \mathscr{S}_{t^-}\}, \tag{8.1}$$

$\mathscr{S}_{t^-}$ being used to denote the set of pages present in main memory immediately before the tth reference. Under the assumptions of the IRM, King (1971) obtains

$$F(\mathrm{LRU}, C, n) = \sum_{q \in Q} \frac{D_1(q) \prod_{i=1}^{C} b_{j(i)}}{\prod_{i=1}^{C-1} [D_i(q) + b_{j(C)}]}, \tag{8.2}$$

where

$$D_{i(q)} = 1 - \sum_{k=i}^{C} b_{j(k)},$$

with Q the set of all C-tuples $q = [j(1), \ldots, j(C)]$, with $j(k) \in \{1, \ldots, n\}$ and $j(k) \neq j(r)$ iff $k \neq r$. Similar expressions exist for most well-known replacement algorithms (King, 1971; Aho *et al.*, 1971).

Unfortunately, this IRM model does not provide a realistic definition for the program localities. The model allows for a distinction between highly referenced pages and others, but gives no information whatsoever on the different localities to which highly referenced pages might belong. This is the reason why, as already observed by Spirn and Denning (1972), expressions such as (8.2) may produce very poor approximations to reality. We shall observe it ourselves with the numerical example that is treated later in this chapter.

8.2.2 ***The Least Recently Used Stack Model*** **(LRUSM)** So far, this model has yielded the most realistic results (Spirn and Denning, 1972). It is based on the notion of *stack distance probability*. All program pages are ordered in a stack according ot their recency of use. The top of the stack is the currently referenced page, the second position is occupied by the next most recently used page, and so on. When a reference is made to the page in stack position i, this page is moved to the top of the stack while pages in position $1, 2, \ldots, i-1$ are pushed down one position, all other positions remaining unchanged; the *stack distance* d_t is then defined as the stack position of the page being referenced at the tth reference.

Clearly, *for a given initial stack*, a reference string $r_1 r_2 \cdots r_t \cdots$ can be equivalently defined by the corresponding stack distance string $d_1 d_2 \cdots d_t \cdots$. The LRUSM model is then based on the following assumptions:

(i) For all t and $1 \leqslant i \leqslant n$,

$$\Pr\{d_t = i\} = a_i,$$

with $a_i > 0$ and $\sum_{i=1}^{n} a_i = 1$.

(ii) For all $t' \neq t''$ and $1 \leqslant i, j \leqslant n$, $\{d_{t'} = i\}$ and $\{d_{t''} = j\}$ are independent events.

Except for the probability distribution $\{a_i\}$ that replaces the distribution $\{b_i\}$, these assumptions are similar to those of the IRM. The advantage of the $\{a_i\}$ over the $\{b_i\}$ distribution is to distinguish between a reference made to a page recently used and thus likely to belong to the current locality and a reference to a page that does not. However, because it is unique and time independent, this $\{a_i\}$ distribution reflects only an average of the locality phenomenon over the whole program; it would fail to identify various localities of different size.

Another deficiency of the LRUSM is its inaptitude to distinguish properly between what we shall call *neighboring localities* and *disjoint localities*. Neighboring localities are localities that differ by a very few pages only; disjoint localities on the contrary have a very few pages in common. Madison and Batson (1976) recently gave experimental evidence that jumps between disjoint localities do exist in actual programs. Although there seems to be no direct experiment proving the existence of neighboring localities, they may conceivably exist as well.

The LRUSM implies that *transitions* always occur between neighboring localities, and makes difficult if not impossible, the modeling of *jumps* between disjoint localities. This may be explained as follows. The set of stack distance probabilities must be (a) biased toward long distances, (b) not biased at all, or (c) biased toward short distances. The first two cases must obviously be interpreted as an absence of localities, and the third case can only be an indication of a presence of neighboring localities. Indeed, in this latter case suppose that the probabilities $\{a_i\}$ are used to produce random numbers and generate references in an attempt to reproduce those same reference strings from which these probabilities have been derived. Then the occasional generation of a single reference to a long-distance page will in general be followed by the generation of many references to short-distance pages. All changes of localities will thus take the form of transition between neighboring localities that differ by one or a very few pages only.

Attempts have been made to overcome this difficulty of modeling jumps between disjoint localities. Spirn and Denning (1972), for example, introduce two sets of stack distance probabilities. One set is biased toward short distances; it is selected most of the time and takes into account slow drifts among neighboring localities. Another set is biased toward long distances; it is selected occasionally only in order to scramble up the stack and mimic sudden jumps between disjoint localities. However, the choice of the probabilities for this set selection can only be very informally justified and remains somewhat arbitrary. Shedler and Tung (1972) proposed a Markov process to generate stack distances, but again the greatest drawback in their approach is that all localities must have the same size.

Except for these deficiencies, the LRUSM model is obviously well suited to study the performances of the LRU replacement algorithm; for example, we have simply

$$F(\mathrm{LRU}, C, n) = \sum_{i=C+1}^{n} a_i.$$

For the other replacement algorithms, however, one has not yet obtained with the LRUSM all the analytical results obtained with the IRM. The few results obtained with the LRUSM can be found in Spirn (1973).

8.2.3 ***Markovian Reference Models*** **(MRM)** These models are based on the knowledge of a matrix **Q**, the element q_{ij} of which is, for any time t, the probability of referencing page j, given that page i was referenced at time $t-1$:

$$q_{ij} = \Pr\{r_t = j \mid r_{t-1} = i\}, \qquad 1 \leqslant i \leqslant n, \quad 1 \leqslant t.$$

We have necessarily that $\sum_{j=1}^{n} q_{ij} = 1$, so that the matrix $\mathbf{Q} = [q_{ij}]$ is stochastic. For a given program this matrix is not difficult to obtain and is clearly based on the simplifying first-order Markovian assumption that for all pages $k, \ldots, s$,

$$\Pr\{r_t = j \mid r_{t-1} = i, r_{t-2} = k, \ldots, r_1 = s\} = \Pr\{r_t = j \mid r_{t-1} = i\}.$$

Such a matrix is not new; it was mentioned by Aho *et al.* (1971) and Mattson *et al.* (1970), and it is similar to the concept of nearness matrix introduced by Hatfield and Gerald (1971) and also to the matrix used by Kral (1968). More recently this matrix has also been used by Franklin and Gupta (1974) and by Glowacki (1974) to calculate the page fault probability of certain replacement algorithms.

These Markovian models are much more adequate than the previous ones to take into account the dynamic change of localities during program execution. Unfortunately, Franklin and Gupta give an algorithm only to evaluate page fault probabilities, but no closed-form expression, while Glowacki's expressions are mathematically almost unmanageable for use in further analysis. So, in both cases little insight can be gained into the influence of the various model parameters. Besides, the MRM has another disadvantage, which is discussed in the next section.

8.2.4 ***Concluding Remarks*** Apart from the models just described, a few others have been proposed in the literature: a review of them can be found in (Courtois and Vantilborgh, 1976). However, these models are, in our opinion, of lesser practical value because they require in one way or another some a priori knowledge of the locality structure of the programs to be analyzed. Such knowledge is, in practice, not required by the models we have mentioned: the probability sets $\{a_i\}$ and $\{b_i\}$, and the Markov matrix **Q** can be easily measured for a given program.

We are left with the rather disappointing conclusion that a model of program behavior is either advantageous from the analytical point of view—and then, as the IRM, poor in its representation of the locality phenomenon—or adequate to model the locality phenomenon—and then, as the MRM, mathematically almost unmanageable. The LRUSM stands as a somewhat arbitrary compromise between these two issues.

Moreover, the IRM and the MRM have one further—and disastrous—shortcoming: even for small sizes of the program page set, their requirements in computer time and space are beyond actual possibilities. As an example let us evaluate approximately the number of multiplications needed to calculate the page fault rate of the LRU replacement algorithm under the hypotheses of the IRM (King, 1971) and of the MRM (Franklin and Gupta, 1974; Glowacki, 1974). King's formula requires $2C$ multiplications for each of the $n!/(n-C)!$ distinct memory states; the algorithm of Franklin and Gupta requires the inversion of $n-1$ matrices of order $n!/(n-C)!$; Glowacki's formula requires the inversion of $C-1$ matrices of order C for each of the

TABLE 8.1
APPROXIMATE NUMBER OF MULTIPLICATIONS NEEDED TO CALCULATE $F(\mathrm{LRU}, C, n)$

Author	General case	$n = 5$	$n = 15$	$n = 25$
King (1971)	$\sum_{C=1}^{n-1} \frac{n!}{(n-C)!} 2C$	1.4×10^3	6.0×10^{13}	1.2×10^{27}
Franklin and Gupta (1974)	$\sum_{C=1}^{n-1} \frac{1}{3}\left(\frac{n!}{(n-C)!}\right)^3$	6.5×10^5	8.4×10^{35}	1.4×10^{75}
Glowacki (1974)	$\sum_{C=1}^{n-1} \frac{n!}{(n-C)!}(C-1)\frac{1}{3}C^3$	8.8×10^3	2.3×10^{16}	2.6×10^{30}

$n!/(n-C)!$ memory states. Table 8.1 sums up these requirements for the general case and for some values of n.

The nearly completely decomposable model we shall propose in the next section attempts to keep the advantages of the previous models and to dispense with this computational shortcoming, while taking into account the locality structure of the programs.

8.3 Nearly Completely Decomposable Model of Program Behavior

8.3.1 When it represents a program that has localities, a Markov matrix such as the matrix $\mathbf{Q}$ introduced in Section 8.2.3 enjoys a special property that so far has not been exploited. Suppose that its rows and columns are arranged so that pages of a same locality have consecutive entries. Then, by definition of a locality, $\mathbf{Q}$ will have along the main diagonal a string of square submatrices of large elements. If we suppose now that all these localities are what we shall call *strictly disjoint*, i.e., that each page belongs to exactly one locality, then these submatrices will be principal submatrices of $\mathbf{Q}$. Moreover, outside these submatrices all elements will be comparatively small as they are probabilities of referencing a page that does not belong to the current locality. Therefore, $\mathbf{Q}$ in this case is *nearly completely decomposable*; it takes the form (1.3), i.e.,

$$\mathbf{Q} = \mathbf{Q}^* + \varepsilon\mathbf{C},$$

where

$$\mathbf{Q}^* = \begin{bmatrix} \mathbf{Q}_1{}^* & & & \\ & \mathbf{Q}_2{}^* & & \\ & & \ddots & \\ & & & \mathbf{Q}_N{}^* \end{bmatrix},$$

with the elements not displayed equal to 0. $\mathbf{Q}^*$ is also a stochastic matrix of order n, completely decomposable into the matrices $\mathbf{Q}_I{}^*$ or order $n(I)$; N is the number of disjoint localities and $n(I)$ the size of locality I. Each matrix $\mathbf{Q}_I{}^*$ is a separate model of program behavior within locality I. The real number ε is defined by (1.4), viz.,

$$\varepsilon = \max_{1 \leqslant I \leqslant N} \left(\max_{1 \leqslant i \leqslant n(I)} \left(\sum_{J \neq I} \sum_{j=1}^{n(J)} q_{i_I j_J} \right) \right),$$

where $q_{i_I j_J}$ denotes the element of $\mathbf{Q}$ at the intersection of the ith row and jth column of $\mathbf{Q}_{IJ}$, which is the submatrix of $\mathbf{Q}$ at the intersection of the Ith set of rows and Jth set of columns. Thus ε is small compared to the elements of $\mathbf{Q}$; *it is the maximum probability of leaving a locality.*

8.3.2 Since $\mathbf{Q}$ is nearly completely decomposable, the Simon–Ando theorems are applicable and our approach consists simply in fully exploiting these theorems.

Let $[y_{i_I}(t)]$ denote a row vector of n elements, where $y_{i_I}(t)$ is the probability that $\mathbf{Q}$ is in state i_I at time t, $1 \leqslant i \leqslant n(I)$, $1 \leqslant I \leqslant N$, $1 \leqslant t$; more precisely, in this case $y_{i_I}(t)$ is the probability of a reference at time t to page i of locality I. We have

$$[y_{i_I}(t+1)] = [y_{i_I}(t)]\mathbf{Q}, \qquad 1 \leqslant t.$$

Let also $[v_{i_I}]$ denote the steady-state vector of $\mathbf{Q}$

$$v_{i_I} = \lim_{t \to \infty} y_{i_I}(t),$$

assuming that all conditions are fulfilled (irreducibility of $\mathbf{Q}$) for this limit to exist and to be unique.

This probability $y_{i_I}(t)$ that page i of locality I is referenced at time t can be considered as the product of

(a) the probability that locality I is referenced at time t, and

(b) the conditional probability that page i is referenced at time t, given that the reference is at t directed to locality I.

The conditional probability (b) is equal to

$$y_{i_I}(t)\left(\sum_{i=1}^{n(I)} y_{i_I}(t)\right)^{-1}. \tag{8.3}$$

Let us then remember the following consequences of the Simon–Ando theorems. In the short-term period, equilibrium states are reached approximately *separately* by each subsystem $\mathbf{Q}_{II}$; and these equilibrium states are approximately *preserved* during the long-term period. More precisely, the probabilities $y_{i_I}(t)$, $1 \leqslant i \leqslant n(I)$ of each subset I vary during the long-term period in roughly the same proportion, keeping approximately the same ratios. Thus, during this long-term period, the conditional probability (8.3) remains approximately equal to its steady-state value

$$v_{i_I}\beta_I^{-1}, \qquad i = 1, \ldots, n(I), \tag{8.4}$$

with

$$\beta_I = \sum_{i=1}^{n(I)} v_{i_I}. \tag{8.5}$$

Moreover, the probability of a transition from locality I to locality J at some time t, i.e., the probability that, some page of I being referenced at time t, a page of J is referenced at time $t+1$ is given by

$$\sum_{i=1}^{n(I)} y_{i_I}(t)\left(\sum_{i=1}^{n(I)} y_{i_I}(t)\right)^{-1} \sum_{j=1}^{n(J)} q_{i_I j_J}.$$

Thus, this probability also remains approximately time independent during the long-term period and equal to, say,

$$p_{IJ} = \beta_I^{-1} \sum_{i=1}^{n(I)} v_{i_I} \sum_{j=1}^{n(J)} q_{i_I j_J}.$$

The stochastic matrix $\mathbf{P} = [p_{IJ}]$ is the transition matrix of an N-state time-independent Markov chain model of interlocality transitions, with holding times τ_I that are geometrically distributed with mean $(1-p_{II})^{-1}$; its steady-state vector is the row vector $[\beta_I]$, $1 \leqslant I \leqslant N$, β_I being the steady-state probability that a reference is directed to locality I, as can be seen from (8.5).

8.3.3 In other words, the distribution of references over the pages of each locality I (i.e., subsystem $\mathbf{Q}_{II}$) approaches during the short-term period, almost independently from references to other localities, a statistical equilibrium defined by the distribution (8.4). These locality equilibriums are approximately maintained in the long-term period when the distribution of references over the different localities approaches the distribution β_I, $I = 1, \ldots, N$. These properties justify

(1) for each locality a separate analysis of its own behavior and characteristics (such as its contribution to the page fault rate) in function of the distribution of references over its pages only; and

(2) a determination of the global characteristics of the whole program in terms of the equilibrium values of the locality characteristics weighted by the distribution β_I.

We are therefore using two types of models. A model of program long-term behavior, i.e., a Markov chain model of the interlocality transitions; and models of locality short-term behavior. This combination of models is supported by the principle of near-complete decomposability, which relates a partitioning of the program state space (here, into localities) to a partitioning of its time behavior into short- and long-term dynamics.

8.3.4 ***Models of Locality Short-Term Behavior*** No specific model is in principle required by this approach to determine the short-term characteristics of each locality. It is not even necessary to use the same model for all the localities of a same program (heterogeneous aggregation, cf. Section 6.3.1).

It is, however, interesting to remark that the approximate time independence of the short-term equilibrium distribution of the references over the pages of a same locality is in agreement with assumption (i) of the IRM (cf. Section 8.2.1). This suggests that, if the second IRM assumption is also verified, an IRM with time-independent page reference probabilities (8.4) can be a satisfactory model of locality short-term equilibrium.

But this second (and stronger) IRM assumption is rigorously verified by a locality I only if $\mathbf{Q}_{II}$ is degenerate and has equal elements in each column (or identical rows): then, clearly, the probability of a reference to a given page of locality I is independent of any previously referenced page of this locality. The IRM is thus an acceptable model of short-term equilibrium for localities with matrices having little disparity in each column. For such localities—likely to exist since matrices $\mathbf{Q}_{II}$ have by definition elements of approximately same order of magnitude—the IRM with its analytical advantages is certainly of primary interest.

Typical cases of localities that violate the second IRM assumption are localities with matrices displaying sublocalities or cyclic classes. Sublocalities and cyclic classes are two opposite cases of a strong dependence of the current reference upon the past; we will verify in the next section that in its evaluation of the page fault rate the IRM is too pessimistic in the first case and too optimistic in the second.

Localities with cyclic classes, or more generally with excessive disparity per column must be treated by other models than the IRM; cycles can be treated by deterministic models (see the treatment of the cyclic matrix $\mathbf{Q}_2{}^*$ in the numerical example of Section 8.5); otherwise the MRM must be used to take into account the dependence on past references.

The presence of sublocalities in a locality can be avoided simply by decomposing the program into localities that are sufficiently small to be indecomposable. At any rate these elementary and indecomposable localities must be small enough to keep the calculations required within feasible bounds; otherwise simulation remains the ultimate solution.

We shall now see how our approach may yield better approximations to the page fault rate and to the distribution of the working-set size than currently available models—at least when the assumption of disjoint localities is verified. Our approach indeed is in a way complementary to the LRUSM model in the sense that the more disjoint the localities, the better it performs. We will deal in Section 8.7 with the case of localities that are not strictly disjoint.

8.4 Page Fault Rate

8.4.1 For a given page replacement algorithm A, a given memory size C available to a given n-page program, we have used $F(A, C, n)$ to denote the stationary probability of a page fault

$$F(A, C, n) = \lim_{t\to\infty} \Pr\{r_t = i \textit{ and } i \notin \mathcal{S}_{t^-} \textit{ and } 1 \leqslant i \leqslant n\},$$

$\mathcal{S}_{t^-}$ being used to denote the set of pages present in main memory immediately before the tth reference.

Let us also define the conditional probability

$$F_I(A, C, n) = \lim_{t\to\infty} \Pr\{r_t = i_I \textit{ and } i_I \notin \mathcal{S}_{t^-} \textit{ and } 1 \leqslant i \leqslant n(I)\},$$

where $F_I(A, C, n)$ is the stationary conditional probability of a page fault *given* that the page being referenced belongs to locality I; the argument n indicates that pages of other localities may be present in main memory upon occurrence of this page fault.

Since β_I is the stationary probability that a page being referenced belongs to locality I, we have

$$F(A, C, n) = \sum_{I=1}^{N} \beta_I F_I(A, C, n). \tag{8.6}$$

8.4.2 Under the assumptions of near-complete decomposability, $F(A, C, n)$ defined by (8.6) can be evaluated as follows. From the argumentation of Section 8.3.3, we know that $F_I(A, C, n)$ can be evaluated separately for each locality I, as an equilibrium characteristic of this locality, and in function of the distribution of references over its pages only. Moreover, a replacement algorithm such as FIFO, LRU, or WORKING SET, upon entry to a new locality I, evacuates from main memory, one by one, those pages of the past localities that obstruct the new locality I; all such pages are evacuated after a number of page faults equal to $\min[C, n(I)]$. Thus, for such replacement algorithms, only the presence in $\mathscr{S}_{t^-}$ of the $n(I)$ pages of locality I need to be considered in the evaluation of $F_I(A, C, n)$ if, with an error of order ε, we neglect the effects of locality transitions. Hence, $F_I(A, C, n)$ can be approximated by the *unconditional* page fault probability, say $F(A, C, n(I))$, of a program of $n(I)$ pages only, with reference probabilities $v_{i_I} \beta_I^{-1}$ and page transition matrix $\mathbf{Q}_I^*$. And with an error of order ε, from (8.6) we obtain

$$F(A, C, n) \approx \sum_{I=1}^{N} \beta_I F(A, C, n(I)). \tag{8.7}$$

8.4.3 For most well-known replacement algorithms $F(A, C, n(I))$ can be calculated by means of derivations similar to those of Aho *et al.* (1971), Franklin and Gupta (1974), Gelenbe (1973), Glowacki (1974), and King (1971). For a locality that verifies the IRM assumptions (cf. Section 8.3.4) and for the LRU replacement algorithm, $F(\text{LRU}, C, n(I))$ can, for example, be obtained by (8.2). If the IRM assumptions are not verified by a locality, Glowacki's or Franklin and Gupta's results can be used to evaluate $F(A, C, n(I))$ or if the arithmetic costs are prohibitive, simulation can be used. Some of these possibilities are illustrated by the numerical example given in the next section.

8.4.4 ***Remark*** As stated above, our approach is especially adequate for algorithms such as FIFO, LRU, and WORKING SET; further elaboration is needed for other replacement algorithms. For RAND, more page faults than $\min(C, n(I))$ are on the average required before pages of the new locality I have completely replaced in main memory the pages of the former one; this average number of additional page faults could, however, be estimated and taken into account. With algorithms such as the partially preloaded algorithms (Gelenbe, 1973), certain pages are permanently encumbering the memory; in this case the above approach would require an evaluation of the amount of memory space effectively available to each locality. These refinements, however, are questionable in view of the uncertain practical interest of these algorithms.

8.4.5 ***Remark*** The approximation (8.7) of $F(A, C, n)$ is a more realistic measure than the page fault probability, say $F'(A, C, n)$, that could be obtained by a *global* IRM, i.e., by a unique IRM applied to the totality of the n pages with page reference probabilities v_{i_I}, $I = 1, \ldots, N$, $i = 1, \ldots, n(I)$.

Indeed, (8.7) reflects the fact that, in accordance with the definition of localities, a program references with probability β_I only $n(I)$ out of its n pages $(1 \leqslant I \leqslant N)$ during sequences of references of average length equal to $(1-p_{II})^{-1}$, the mean holding time of the locality $\mathbf{Q}_{II}$. In the global IRM, on the contrary, all n pages are considered as competing for main memory during any such sequence.

This can be explained somewhat more precisely. Let us use $t(I)+l(I)$ to denote the length of the reference substring that corresponds to an execution of locality I; $t(I)$ is the length of the transition period during which $\min(C, n(I))$ page faults occur; $l(I)$ is the length of the remaining period during which any new missing page of locality I replaces a page also of locality I. Our model ignores the $t(I)$ first references and the corresponding $\min(C, n(I))$ page faults; in the global IRM, on the other hand, each locality I is executed in an actual memory space, say $C'(C' < C)$, which is reduced by the space occupied by the competing highly referenced pages of other localities. Consequently, while in reality the number of page faults during the execution of locality I should be given by

$$\min(C, n(I)) + l(I)\, F(A, C, n(I)),$$

our approach counts

$$(t(I)+l(I))\, F(A, C, n(I)),$$

and the global IRM counts

$$(t(I)+l(I))\, F(A, C', n(I)).$$

The resulting error for the page fault probability in our model is

$$(t(I)+l(I))^{-1}\{t(I)\, F(A, C, n(I)) - \min(C, n(I))\}$$

and in the global IRM

$$F(A, C', n(I)) - F(A, (C, nI)) + (t(I)+l(I))^{-1} \times \{t(I)\, F(A, C, n(I)) - \min(C, n(I))\}. \tag{8.8}$$

Since $(t(I)+l(I))$, $1 \leqslant I \leqslant N$, is proportional to ε^{-1} and since $C' < C$ generally implies that $F(A, C', n(I)) > F(A, C, n(I))$, we conclude that the error on the page fault probability is with our approach proportional to ε, while it is independent of ε with the global IRM. Moreoever, in this latter case (8.8) shows that the error will always be an error by excess. These considerations are corroborated by the numerical example dealt with in the next section. Spirn and Denning (1972) also give experimental evidence for the too pessimistic evaluation of the page fault probability by the global IRM.

8.4.6 ***Remark*** For values of $C \geqslant \max_{1 \leqslant I \leqslant N}(n(I))$, the approximation (8.7) of $F(A, C, n)$ is identically zero since for any particular locality I, $F(A, C, n(I))$ is zero when $C \geqslant n(I)$. This causes no more than an error of order ε^2. Indeed, for $C \geqslant \max_{1 \leqslant I \leqslant N}(n(I))$, page faults occur only at locality transitions, and consequently the page fault rate is smaller than

$$\varepsilon \sum_{I=1}^{N} \beta_I n(I) \left(t(I) + l(I)\right)^{-1} \approx \varepsilon^2 \sum_{I=1}^{N} \beta_I n(I). \tag{8.9}$$

In the same range of C values, the global IRM yields much higher values than (8.9) for the page fault rate.

8.5 Numerical Example

8.5.1 The purpose of this section is to illustrate our approach and to compare its performance with other models. We chose for this comparison the evaluation of the page fault probability of the LRU replacement algorithm; all models deal with this parameter, which is one of the most interesting in practice.

8.5.2 We assume that the 15×15 page transition matrix $\mathbf{Q}$, displayed in Fig. 8.1, is given. It is possible to derive from this matrix a completely decomposable matrix $\mathbf{Q}^*$ consisting of three strictly disjoint localities:

$$\mathbf{Q} = \mathbf{Q}^* + \varepsilon \mathbf{C},$$

where

$$\varepsilon = 0.03, \qquad N = 3, \qquad n(1) = 6, \qquad n(2) = 5, \qquad n(3) = 4$$

$$\mathbf{Q} = \left[\begin{array}{cccccc|ccccc|cccc}
0.100 & 0.200 & 0.290 & 0.090 & 0.200 & 0.100 & 0 & 0 & 0.005 & 0.005 & 0 & 0 & 0.008 & 0.002 & 0 \\
0.148 & 0.150 & 0.150 & 0.142 & 0.150 & 0.240 & 0.003 & 0.003 & 0 & 0 & 0.004 & 0.002 & 0.003 & 0 & 0.005 \\
0.280 & 0 & 0.300 & 0 & 0.300 & 0.100 & 0.010 & 0 & 0 & 0 & 0.002 & 0.006 & 0 & 0 & 0.002 \\
0.100 & 0.280 & 0 & 0.300 & 0 & 0.290 & 0.010 & 0 & 0.010 & 0 & 0 & 0 & 0.010 & 0 & 0 \\
0.380 & 0.300 & 0.050 & 0.050 & 0.050 & 0.150 & 0 & 0 & 0.005 & 0 & 0.005 & 0.005 & 0.005 & 0 & 0 \\
0.100 & 0.100 & 0.100 & 0.100 & 0.490 & 0.100 & 0 & 0 & 0 & 0 & 0.005 & 0.005 & 0 & 0 & 0 \\
\hline
0 & 0 & 0.005 & 0 & 0.005 & 0 & 0 & 0.980 & 0 & 0 & 0 & 0 & 0 & 0.005 & 0.005 \\
0 & 0.005 & 0 & 0.005 & 0 & 0 & 0 & 0 & 0.980 & 0 & 0 & 0 & 0.005 & 0 & 0.005 \\
0.005 & 0 & 0.005 & 0 & 0 & 0 & 0 & 0 & 0 & 0.980 & 0 & 0 & 0.005 & 0.005 & 0 \\
0 & 0.005 & 0 & 0 & 0.005 & 0 & 0 & 0 & 0 & 0 & 0.980 & 0 & 0.005 & 0 & 0.005 \\
0.005 & 0 & 0.005 & 0 & 0 & 0.005 & 0.980 & 0 & 0 & 0 & 0 & 0.005 & 0 & 0 & 0 \\
\hline
0.005 & 0 & 0 & 0.005 & 0 & 0 & 0.005 & 0 & 0 & 0.005 & 0 & 0.050 & 0.400 & 0.100 & 0.430 \\
0 & 0.005 & 0 & 0 & 0.005 & 0 & 0 & 0.005 & 0 & 0 & 0.005 & 0.680 & 0.050 & 0.250 & 0 \\
0 & 0 & 0.005 & 0 & 0 & 0 & 0 & 0 & 0.005 & 0 & 0 & 0.590 & 0 & 0.100 & 0.300 \\
0 & 0 & 0 & 0.010 & 0 & 0.010 & 0 & 0 & 0 & 0 & 0 & 0.050 & 0.300 & 0.580 & 0.050
\end{array}\right]$$

FIGURE 8.1 Page transition matrix.

and

$$\mathbf{Q}^* = \begin{bmatrix} \mathbf{Q}_1{}^* & & \\ & \mathbf{Q}_2{}^* & \\ & & \mathbf{Q}_3{}^* \end{bmatrix}$$

with

$$\mathbf{Q}_1{}^* = \begin{bmatrix} 0.10 & 0.20 & 0.30 & 0.10 & 0.20 & 0.10 \\ 0.15 & 0.15 & 0.15 & 0.15 & 0.15 & 0.25 \\ 0.30 & 0.00 & 0.30 & 0.00 & 0.30 & 0.10 \\ 0.10 & 0.30 & 0.00 & 0.30 & 0.00 & 0.30 \\ 0.40 & 0.30 & 0.05 & 0.05 & 0.05 & 0.15 \\ 0.10 & 0.10 & 0.10 & 0.10 & 0.50 & 0.10 \end{bmatrix},$$

$$\mathbf{Q}_2{}^* = \begin{bmatrix} 0 & 1 & 0 & 0 & 0 \\ 0 & 0 & 1 & 0 & 0 \\ 0 & 0 & 0 & 1 & 0 \\ 0 & 0 & 0 & 0 & 1 \\ 1 & 0 & 0 & 0 & 0 \end{bmatrix},$$

$$\mathbf{Q}_3{}^* = \begin{bmatrix} 0.05 & 0.40 & 0.10 & 0.45 \\ 0.70 & 0.05 & 0.25 & 0.00 \\ 0.60 & 0.00 & 0.10 & 0.30 \\ 0.05 & 0.30 & 0.60 & 0.05 \end{bmatrix}.$$

Let us remark that the matrix $\mathbf{Q}^*$ is obtained from $\mathbf{Q}$ in a way that increases the indecomposability of the aggregates $\mathbf{Q}_I{}^*$, $I = 1, 2, 3$, in accordance with the discussion of Section 2.2.2. The page fault probability of the LRU replacement algorithm is obtained by (8.6), viz.,

$$F(\mathrm{LRU}, C, 15) = \sum_{I=1}^{3} \beta_I F(\mathrm{LRU}, C, n(I)). \tag{8.10}$$

For illustrative purposes we have chosen the matrices $\mathbf{Q}_1{}^*$, $\mathbf{Q}_2{}^*$, and $\mathbf{Q}_3{}^*$ in such a way that a same model cannot be used for the three localities.

In locality 1, we use an IRM with reference probabilities given by the equilibrium probability vector of $\mathbf{Q}_1{}^*$, viz.,

$$[v_{i_1}^*] = [0.202 \quad 0.174 \quad 0.161 \quad 0.103 \quad 0.203 \quad 0.157].$$

The page fault probabilities obtained by King's formula (8.2) are given in Table 8.2; this table gives also, by way of comparison, the corresponding values yielded by Glowacki's MRM approach. The good match between

TABLE 8.2

$F(\mathrm{LRU}, C, n(1))$ BY IRM AND MRM

C	$F(\mathrm{LRU}, C, n(1))$ by IRM	$F(\mathrm{LRU}, C, n(1))$ by MRM
1	0.827	0.849
2	0.654	0.655
3	0.484	0.469
4	0.316	0.292
5	0.152	0.130
$\geqslant 6$	0	0

these two sets of values shows that the IRM can remain satisfactory even for matrices such as $\mathbf{Q}_1^*$ that depart substantially from the ideal situation of identical rows mentioned in Section 8.3.4.

In locality 2, the behavior of the program is cyclic, and the matrix $\mathbf{Q}_2^*$ models a deterministic cycle of length 5; we have simply

$$F(\mathrm{LRU}, C, n(22)) = \begin{cases} 1 & \text{if } 1 \leqslant C \leqslant 4, \\ 0 & \text{otherwise.} \end{cases}$$

In locality 3 the rows of matrix $\mathbf{Q}_3^*$ are too dissimilar, so that the exact MRM must be used for determining $F(\mathrm{LRU}, C, n(3))$. With Glowacki's formula we obtain the results shown in Table 8.3. The quantities β_1, β_2, and β_3 needed to calculate $F(\mathrm{LRU}, C, 15)$ by (8.6) are obtained by (8.5), viz.,

$$\beta_1 = 0.362, \qquad \beta_2 = 0.301, \qquad \beta_3 = 0.337.$$

The values of $F(\mathrm{LRU}, C, 15)$ obtained by (8.6) are shown in Table 8.4.

8.5.3 We would have liked to compare our results of Table 8.4 with similar results *calculated* for the same matrix $\mathbf{Q}$ by means of the other models discussed in Section 8.2, but such calculations are unfortunately prohibitive. For

TABLE 8.3

$F(\mathrm{LRU}, C, n(3))$ BY MRM

C	$F(\mathrm{LRU}, C, n(3))$
1	0.9377
2	0.6407
3	0.2711
$\geqslant 4$	0

TABLE 8.4

$F(\mathrm{LRU}, C, 15)$ BY NEAR-COMPLETE DECOMPOSABILITY

C	$F(\mathrm{LRU}, C, 15)$
1	0.916
2	0.754
3	0.568
4	0.416
5	0.055
$\geqslant 6$	0

the matrix **Q**, approximately 6.0×10^{13} multiplications are required by King's formula, 8.4×10^{35} by Franklin and Gupta's algorithm, and 2.3×10^{16} by Glowacki's approach. Our approach requires the inversion of $N+1$ matrices of order $N, n(1), \ldots, n(N)$, respectively, and for each of the $n(I)!/(n(I)-C)!$ memory states, $1 \leqslant C \leqslant n(I)$, $1 \leqslant I \leqslant N$, of each locality I a number M_I of multiplications that is a function of the model used in this locality (see Table 8.1); the approximate number of multiplications is thus equal to

$$\frac{1}{3}\left(N^3 + \sum_{I=1}^{N} n(I)^3\right) + \sum_{I=1}^{N} \sum_{C=1}^{n(I)-1} \frac{n(I)!}{(n(I)-C)!} M_I,$$

which in our example equals 1.3×10^4. This shows the important computational advantages of a nearly completely decomposable model. It is, of course, achieved at the expense of an approximation, the bounds of which, however, remain known.

8.5.4 For our comparison, we had to resort to simulation. We evaluated by simulation the global IRM and the MRM approach. For the global IRM we calculated the LRU stack distance frequencies (see Section 8.2.2) of a sequence of independent page references randomly generated according to the distribution defined by the steady-state vector $[v_{i_I}]$ of the matrix **Q**. This IRM simulation was carried out until there was a sufficient agreement between this vector $[v_{i_I}]$ and the frequencies of the references actually observed (the maximal relative error was 5%). Similarly for the MRM, the LRU stack distances were observed on a string of references generated in accordance with the transition probabilities of matrix **Q**. This MRM simulation was continued until

(i) the distribution of actual references over the three localities was in sufficient agreement with the distribution $[\beta_I]$; the simulation gave [0.328 0.325 0.347] instead of $[\beta_I] = [0.362\ 0.301\ 0.337]$; and

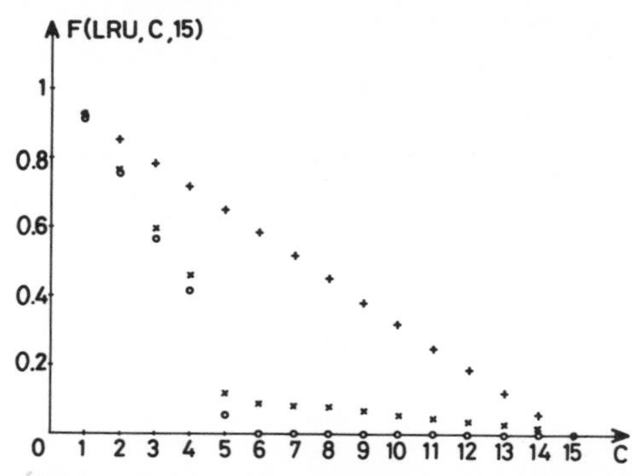

FIGURE 8.2 $F(\mathrm{LRU}, C, 15)$ (×, MRM, simulated; O, near-complete decomposability, calculated; +, global IRM, simulated).

(ii) the local equilibrium $[v_{i_I}^*]$ was sufficiently approached in each locality; the maximal relative errors in the three localities were 4, 1, and 3%, respectively.[1]

These results are displayed in Fig. 8.2. Taking the MRM as a basis of comparison, one observes that the values yielded by our approach differ by an amount of order ε from the estimated MRM values, while the estimated values for the global IRM are strikingly different. These differences must be interpreted here as an absolute lack of validity of the global IRM assumptions. The estimated page fault rate for the global IRM varies indeed almost linearly with the memory size, which shows that this model fails to take into account the important part played by the localities in the model of program behavior defined by **Q**.

8.6 Working-Set Size Distribution

8.6.1 In the case of disjoint localities, our method also yields better approximations to the working-set size distribution than currently available models.

The *working set* $W(t, T, n)$ of a program that has n distinct page is defined (Denning, 1968c), at the tth reference, as being the set of the pages that were referenced during the last T references. T is called the window; the size of the working set, denoted $w(t, T, n)$, is the number of distinct pages contained in it. This concept is intended to provide the store management strategy with a means of automatic and dynamic identification of the localities of the program in execution. The page replacement algorithm based on this concept is then restricted to choose its victim among the pages that do not belong to the current working set. Ideally the current working set should contain the

[1] These halting conditions are obviously determined by the near-complete decomposable property of our model; this indicates that this property might also be of some general use in simulation experiments.

current locality only. The choice of the window T is obviously critical in this connection: if the window is too large or too small, the working set will contain more than or less than the current locality, respectively. Although it will always remain a difficult issue, the knowledge of the working set size distribution for different choices of T may help in finding an optimal window for a given family of programs.

8.6.2 Let

$$g(k,T,n) = \lim_{t\to\infty} \Pr\{w(t,T,n) = k\}, \qquad 1 \leqslant k \leqslant n,$$

be the *stationary probability density function of the working-set size* for a window T. Let us also define the conditional stationary probability density function

$$g_I(k,T,n) = \lim_{t\to\infty} \Pr\{w(t,T,n) = k \mid r_t = i_I,\ 1 \leqslant i \leqslant n(I)\}, 1 \leqslant k \leqslant n;$$

$g_I(k,T,n)$ is the stationary probability that the working set has size k *given* that the current page being referenced belongs to locality I. Since β_I is the steady-state probability that a page being referenced belongs to locality I, we have

$$g(k,T,n) = \sum_{I=1}^{N} \beta_I g_I(k,T,n). \tag{8.11}$$

Thanks to the properties of near-complete decomposability, the functions $g_I(k,T,n)$, $1 \leqslant I \leqslant N$, can be approximated in the following way. Suppose that $T \leqslant \min_{1\leqslant I\leqslant N}(\tau_I)$; this choice of a window smaller than the minimum of the average locality holding times τ_I will be discussed later. In this case, the working set will contain pages of different localities at locality transitions only. By definition, such transitions occur infrequently, i.e., with a probability which, at each reference, is at most equal to ε. Thus, upon a reference to a page of locality I, the working set will contain pages of this locality I *only*, with a probability at least equal to $(1-\varepsilon)^T$.

Besides, page references to this locality reach, almost independently of other localities, the short-term equilibrium probability distribution $\{v_{i_I}\beta_I^{-1}\}$, $1 \leqslant i \leqslant n(I)$. Thus the *conditional* probability density function $g_I(k,T,n)$ can be evaluated, with an error of order ε, as the *unconditional* probability density function of the working-set size for a program with $n(I)$ pages only and with reference probabilities $v_{i_I}\beta_I^{-1}$. Moreoever, if the IRM assumption (ii) is also valid for this locality, the IRM formulas given by Vantilborgh (1974) are applicable, and for $1 \leqslant k \leqslant n(I)$, we obtain

$$g_I(k,T,n) \approx g(k,T,n(I)) = \sum_{h=1}^{k} (-1)^{k-h} \binom{n(I)-h}{n(I)-k} D_I(h,T,n(I)), \tag{8.12}$$

where

$$D_I(h,T,n(I)) = \sum_{1 \leqslant i(1) < \cdots < i(h) \leqslant n(I)} (v_{i(1)_I} + \cdots + v_{i(h)_I})^T \beta_I^{-T}.$$

It also follows from (8.11) that the *mean working set size* $\bar{w}(T,n)$ is given by

$$\bar{w}(T,n) = \sum_{I=1}^{N} \beta_I \bar{w}_I(T,n), \tag{8.13}$$

where $\bar{w}_I(T,n)$ is the conditional mean working-set size *given* that the current page being referenced belongs to locality I. Again, under the IRM assumptions, $\bar{w}_I(T,n)$ can be approximated with an accuracy of order ε by (Denning and Schwartz, 1972)

$$\bar{w}_I(T,n) \approx \bar{w}(T,n(I)) = n(I) - \sum_{i=1}^{n(I)} (1 - v_{i_I} \beta_I^{-1})^T. \tag{8.14}$$

The *variance* $\sigma^2(T,n)$ of $g(k,T,n)$ can be obtained as follows:

$$\begin{aligned}
\sigma^2(T,n) &= \sum_{k=1}^{n} k^2 g(k,T,n) - \bar{w}^2(T,n) \\
&= \sum_{I=1}^{N} \beta_I \sum_{k=1}^{n} k^2 g_I(k,T,n) - \bar{w}^2(T,n) \\
&= \sum_{I=1}^{N} \beta_I \sigma_I^2(T,n) + \sum_{I=1}^{N} \beta_I (\bar{w}_I^2(T,n)) - \left(\sum_{I=1}^{N} \beta_I \bar{w}_I(T,n) \right)^2.
\end{aligned}$$

Multiplying each term of the second summation by $1 = \sum_{I=1}^{N} \beta_I$, and working out the square of the third summation, we obtain

$$\sigma^2(T,n) = \sum_{I=1}^{N} \beta_I \sigma_I^2(T,n) + \tfrac{1}{2} \sum_{I=1}^{N} \sum_{J=1}^{N} \beta_I \beta_J (\bar{w}_I(T,n) - \bar{w}_J(T,n))^2. \tag{8.15}$$

In this expression $\sigma_I^2(T,n)$ is the variance of $g_I(k,T,n)$; when an IRM can be used within locality I, then $g_I(k,T,n)$ is approximated by (8.12), and $\sigma_I^2(T,n)$ by (Vantilborgh, 1974)

$$\begin{aligned}
\sigma^2(T,n(I)) = \sum_{i'=1}^{n(I)} \sum_{i''=1,\, i'' \neq i'}^{n(I)} & (1 - (v_{i_I'} + v_{i_I''}) \beta_I^{-1})^T \\
& - \{n(I) - \bar{w}(T,n(I))\}\{n(I) - \bar{w}(T,n(I)) - 1\}.
\end{aligned}$$

$\bar{w}_I(T,n)$ is then approximated by (8.14).

For an indecomposable locality that does not verify the IRM assumptions, one must resort to simulation for determining $g(k,T,n(I))$, $\bar{w}(T,n(I))$, and $\sigma^2(T,n(I))$. Closed-form expressions for these quantities are not yet available under MRM assumptions.

8.6.3 ***Normality*** Up to now, theory and experiment have given contradictory results on the issue of whether the working-set size is approximately normally distributed or not. As argued by Denning and Schwartz (1972), one can theoretically expect this normality. This has been explicitly proved for the IRM and the SLRUSM: Vantilborgh (1974) has derived the exact working-set size distribution under the IRM hypotheses and showed that this distribution is approximately normal in the worst case; Coffman and Ryan (1972) investigated the normal approximation with a Monte Carlo simulation under the SLRUSM hypotheses and found this approximation excellent. Experiments, on the contrary, do not always support this normality (Bryant, 1975; Rodriguez-Rosell, 1973).

This apparent contradiction disappears in our model. The normality of the working-set size has been rigorously proved by Vantilborgh (1974) for programs that verify the IRM assumptions. We indicated in Section 8.3.4 how these assumptions are likely to be verified at locality level; for such localities, the function $g(k,T,n(I))$ is thus necessarily approximately normal. The distribution $g(k,T,n)$, on the other hand, takes explicitly into account the distinct sizes of the various localities, and this may lead to a multimodal distribution function. It is precisely this multimodality that was observed by Bryant (1975) and Rodriguez-Rosell (1973). The function $g(k,T,n)$ will exhibit the approximately normal behavior only if all functions $g_I(k,T,n)$ are approximately normal with approximately the same mean. This is likely to be verified when all localities have about the same size, and this explains the approximately normal behavior obtained in the IRM and in the SLRUSM; for the same reason, though it has not yet been investigated, the same behavior can be expected in the VSLM.

This discussion on normality leads to another observation for programs that have localities of different sizes. A good choice for the parameter T of a working set is, as suggested earlier, around a value T_0 such that for almost all times t, $W(t,T_0,n)$ contains one and only one locality: T_0 must be such that

$$\max_{1 \leqslant I \leqslant N} (T_I) \approx T_0 < \min_{1 \leqslant I \leqslant N} (\tau_I), \tag{8.16}$$

where T_I is used to denote the mean time interval needed to refer *at least once* all pages of locality I. The value T_0 that satisfies (8.16) is likely to exist for a given program, since by definition of a locality,

$$T_I \ll \tau_I, \qquad \text{for each} \quad I.$$

Thus, if the localities have different sizes, T_0 *is also a value for which the deviation from normality of the distribution* $g(k,T,n)$ *will be maximum*. This observation may help in the determination of T_0 from $g(k,T,n)$.

8.6.4 ***Remark*** For large values of T the limiting behavior of both $g(k,T,n)$ and $\bar{w}(T,n)$, when $g_I(k,T,n)$ and $\bar{w}_I(T,n)$ are approximated by (8.12) and (8.14), respectively, no longer corresponds to what might be expected in reality. As $T \to \infty$, all n pages tend to be with probability 1 in the working set; thus, the working-set size distribution should approach a ray of length 1 at $k = n$, and the mean working-set size should approach n. This is clearly not true for our approximation of $g(k,T,n)$ and $\bar{w}(T,n)$. As $T \to \infty$, each $g(k,T,n(I))$, as defined by (8.12), approaches a ray of length 1 at $k = n(I)$ (and this is true independently of the model of program behavior used in locality I); and it is also clear from (8.13) and (8.14) that

$$\lim_{T \to \infty} \bar{w}(T,n) = \sum_{I=1}^{N} \beta_I n(I).$$

In other words, as T increases, our approximation of $\bar{w}(T,n)$ approaches the *mean locality size* and not n.

Such deviations from the expected behavior are not critical; they occur for large values of T that are beyond the range of interest. For all values of T around T_0, $g(k,T,n)$ and $\bar{w}(T,n)$ defined by expressions (8.11) and (8.13) behave correctly since most of the time at most one locality is contained in the working set. In fact, these expressions are more accurate in this range than those which would be obtained in a single IRM applied to the totality of the n pages of the program. Indeed, for $T = T_0$, (8.14) yields $\bar{w}(T_0, n(I)) \approx n(I)$, and by substitution in (8.13), we obtain

$$\bar{w}(T,n) \approx \sum_{I=1}^{N} \beta_I n(I),$$

which is the expected result. With $\bar{w}'(T,n)$ denoting the mean working-set size in the global IRM mentioned above, Courtois and Vantilborgh (1976) have shown that $\bar{w}'(T_0,n) > \bar{w}(T_0,n)$; in other words, the global IRM over-estimates the working-set size. A similar observation had already been made by Spirn and Denning (1972).

8.6.5 ***Working-Set Replacement Policy*** When the page to be replaced is always chosen among those which do not belong to the current working set, an upper bound of the page fault probability is given by the *missing page probability*

$$M(T,n) = \lim_{t \to \infty} P\{r_t \notin W(t-1, T, n)\}.$$

With the same argument we have used time and again, we obtain

$$M(T,n) \approx \sum_{I=1}^{N} \beta_I M(T, n(I)). \tag{8.17}$$

In this case it is easy to prove that, if *all* localities verify the IRM assumptions, $M'(T,n)$ yielded by a global IRM is for all $T > M(T,n)$ yielded by (8.17). Indeed, we have (Denning and Schwartz, 1972)

$$M'(T,n) = \sum_{I=1}^{N} \sum_{i=1}^{n(I)} v_{i_I}(1-v_{i_I})^T,$$

and on the other hand,

$$\begin{aligned} M(T,n) &= \sum_{I=1}^{N} \beta_I \sum_{i=1}^{n(I)} v_{i_I}\beta_I^{-1}(1-v_{i_I}\beta_I^{-1})^T \\ &= \sum_{I=1}^{N} \sum_{i=1}^{n(I)} v_{i_I}(1-v_{i_I}\beta_I^{-1})^T. \end{aligned}$$

Due to $v_{i_I} < \beta_I = \sum_{i=1}^{n(I)} v_{i_I} < 1$, the inequality $M(T,n) < M'(T,n)$ is verified.

8.7 Not Strictly Disjoint Localities

When localities are not strictly disjoint and overlap each other, the page transition matrix $\mathbf{Q}$ is, strictly speaking, no longer nearly completely decomposable and the formalism of Section 8.3 is no longer directly applicable. However, the philosophy behind this formalism remains valid: *if the locality transition matrix* $\mathbf{P}$ *is available*, its steady-state locality reference probability vector $[\beta_I]$ can be calculated and the following stochastic averages can still be defined:

$$\begin{aligned} g(k,T,n) &\approx \sum_{I=1}^{N} \beta_I g(k,T,n(I)), \\ \bar{w}(T,n) &\approx \sum_{I=1}^{N} \beta_I \bar{w}(T,n(I)), \qquad (8.18) \\ F(A,C,n) &\approx \sum_{I=1}^{N} \beta_I F(A,C,n(I)]. \end{aligned}$$

As said earlier, there is no restriction on the type of method that can be used to evaluate the locality-dependent quantities $g(k,T,n(I))$, $\bar{w}(T,n(I))$, $F(A,C,n(I)),\ldots$. The main problem that remains is to determine the inter-locality transition matrix $\mathbf{P}$. If the localities have only "a few" pages in common, it remains possible to obtain $\mathbf{P}$ from $\mathbf{Q}$. Let us show this in the particular case when localities do overlap only pairwise, i.e., when localities have a few pages in common with at most two other localities, and each page belongs to at most two localities. Rows and columns of $\mathbf{Q}$ can then be arranged so that large elements remain grouped around the diagonal into square submatrices which have only a few of their entries in common. In Fig. 8.3, for

$$
\mathbf{Q} = \begin{vmatrix}
\times & \times & \times & \times & \times & \bullet & \bullet & \bullet & \bullet \\
\times & \times & \times & \times & \times & \bullet & \bullet & \bullet & \bullet \\
\times & \times & \times & \times & \times & \bullet & \bullet & \bullet & \bullet \\
\times & \times & \times & \times & \times & \times & \times & \bullet & \bullet \\
\times & \times & \times & \times & \times & \times & \times & \bullet & \bullet \\
\bullet & \bullet & \bullet & \times & \times & \times & \times & \bullet & \bullet \\
\bullet & \bullet & \bullet & \times & \times & \times & \times & \times & \times \\
\bullet & \bullet & \bullet & \bullet & \bullet & \bullet & \times & \times & \times \\
\bullet & \bullet & \bullet & \bullet & \bullet & \bullet & \times & \times & \times
\end{vmatrix}
$$

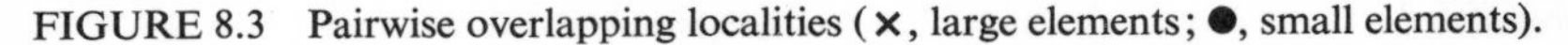

FIGURE 8.3 Pairwise overlapping localities ($\times$, large elements; $\bullet$, small elements).

example, keeping the same index notation, the following states are identical: $4_1 \equiv 1_2$, $5_1 \equiv 2_2$, $4_2 \equiv 1_3$. With $[v_{i_I}]$, $1 \leqslant i \leqslant n(I)$, $1 \leqslant I \leqslant N$, denoting the steady-state vector of **Q**, the elements of **P** are then defined as follows:

(i) for $|I-J| > 1$, by

$$p_{IJ} = \left(\sum_{i=1}^{n(I)} v_{i_I}\right)^{-1} \sum_{i=1}^{n(I)} v_{i_I} \sum_{j=1}^{n(J)} q_{i_I j_J};$$

(ii) for $J = I+1$, by

$$p_{I,I+1} = \left(\sum_{i=1}^{n(I)} v_{i_I}\right)^{-1} \sum_{i=1}^{n(I)} v_{i_I} \sum_{\cdot\, = n(I)_I+1}^{n(I+1)_{I+1}} q_{i_I \cdot};$$

(iii) for $J = I-1$, by

$$p_{I,I-1} = \left(\sum_{i=1}^{n(I)} v_{i_I}\right)^{-1} \sum_{i=1}^{n(I)} v_{i_I} \sum_{\cdot\, = 1_{I-1}}^{1_I - 1} q_{i_I \cdot};$$

(iv) for $J = I$, by

$$p_{II} = 1 - \sum_{J=1, J \neq I}^{N} p_{IJ}.$$

Expressions (ii) and (iii) are based on the assumption that a program is considered as not leaving a locality when it references a page of this locality shared by another. This technique remains applicable, but loses its formal tractability when localities, while remaining sufficiently disjoint, overlap more than pairwise.

When localities have many pages in common (neighboring localities), the matrix **Q** gets completely filled up with large elements, and no information can be gained from it. The long-term probability β_I of referencing in locality I, used in (8.18), should in this case be obtained on the basis of more accurate assumptions than those of a first-order Markov chain. Note also that the more neighboring localities are, the more likely **Q** is to verify the IRM assumptions.

8.8 Conclusion

Existing models of dynamic program behavior may present important discrepancies with reality due to their inability to distinguish and evaluate separately the various localities of a program.

When localities are disjoint, this can be done by analyzing each of them separately as an aggregate of a nearly completely decomposable system; the global behavior of the whole program can then be estimated by combining the results of these separate locality analyses in the framework of a Markovian model of interlocality transitions. An advantage of this approach is that the assumptions of the independent reference model are more likely to be verified at the locality level than at the program level; the great mathematical tractability of the independent reference model can thus be exploited in the local behaviors of the program.

Problems of locality identification remain when localities overlap one another. Although it fails to provide formal results in this case, the nearly completely decomposable approach still, albeit informally, indicates the lines along which to proceed.

An additional advantage of this approach is its computational feasibility. Contrary to other models, its computing time and space requirements remain for practical program sizes within feasible limits. This is achieved at the expense of an approximation, the bounds of which, however, are known.

Let us finally remark that program parameters other than the working-set size distribution and the page fault rate could be evaluated by our approach; recently, Denning and Kahn (1975) used a similar, but experimental approach to reproduce known properties of empirical lifetime functions.

CHAPTER IX

Instabilities and Saturation in Multiprogramming Systems

In this chapter we build and give a detailed analysis of an aggregative model of a typical computing system. Our purpose is twofold. First, our investigation will help us discover and quantify specific properties of the dynamic behavior of the systems captured by our model. Second, we attempt to give the reader an opportunity to assess the pros and cons of our approach in a practical case of system analysis. In this respect we have tried to consider a sufficiently complicated model for all the basic features of real systems to be present; we have also tried to make explicit all the simplifying assumptions required by the analysis.

A set of hardware- and software-dependent parameters will be defined as the model of a classical multiprogramming time-sharing system with a virtual paging memory. Conditions for this model to be nearly completely decomposable will be defined in terms of these parameters and in terms of a stochastic characterization of the work load of the system. The formulation of these conditions makes their prior verification possible and gives an assessment of the precision that may be achieved using an aggregative model.

Page traffic between primary and secondary memory is studied as the internal traffic of a resource *aggregate*. This analysis focuses on the respective influence on processor usage of (i) the user program-paging activity, (ii) the degree of multiprogramming, and (iii) the length of a so-called execution interval, viz., the amount of processor time a user program is allowed to consume before losing the pages it has accumulated in primary memory.

The entire system is then regarded at a higher level of analysis as a finite set of N active user terminals supplying tasks for this aggregate. The congestion and the response time of the system are studied as the characteristics of a system $\mathscr{M}_2(N)$. The concept of *stable* and *unstable* congestions is introduced as the consequence of the peculiar dependency of the aggregate service rate on the system congestion. Likewise, work loads that *saturate* the system are defined. Finally, the evaluation of the system response time reveals that these concepts of stability and saturation relate the phenomenon known as *thrashing* to parameters defining the work load in addition to those defining the page traffic between primary and secondary memory.

Some of the basic results presented in this chapter appeared in Courtois (1975a) and in Courtois and Vantilborgh (1975), where a simpler model is studied that does not take into account the part played by the execution interval.

9.1 System Model

A schematic representation of our model is given in Fig. 9.1. Three types of *functions* are essentially ensured:

(1) A finite number N of *active* user terminals generate random requests for program execution.

(2) User programs are executed on a multiprogrammed basis in a primary memory M_0 containing c_0 page frames.

(3) Pages that cannot be contained in M_0 are swapped on a fixed head per track rotational secondary memory M_1; T_{rot} will designate the duration of a rotation of this memory.

These functions are supposed to comply with the following *strategies*:

9.1.1 A terminal cannot generate a request for program execution before the previous request issued from the same terminal has been served and completed. In other words, there may exist a maximum of one program in execution per terminal in the system at a time.

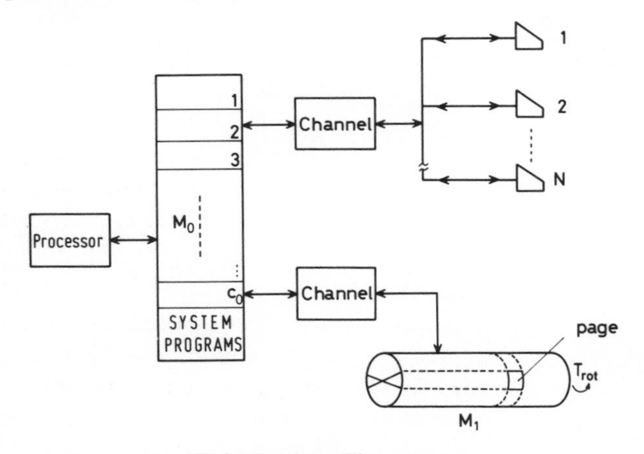

FIGURE 9.1 The system.

9.1.2 User programs are loaded into M_0 from M_1, where all pages are supposed to be initially located, on a *page-on-demand strategy*. An upper limit will be imposed upon the number of page frames each user program may occupy in M_0. Pages of a same user program superimpose each other whenever the number of distinct pages required in M_0 by this program exceeds this upper limit. A *page fault* is said to occur whenever a page is located in M_1 at the time it is referenced.

9.1.3 ***Multiprogramming*** When it is present in M_0, a user program is in one of three states:

ready	demanding but not receiving the control of the processor;
running	having the control of the processor;
suspended	waiting for a page transfer between M_1 and M_0 to be completed.

Multiprogramming means that user programs are concurrently executed in M_0 in order to maintain the processor busy as long as not all multiprogrammed user programs are suspended. We assume that a maximum number J_{max}, $0 < J_{max} \leqslant N$, of user programs may at most be concurrently executed in M_0. Further requests of program executions are queued until one of the J_{max} programs ceases to be multiprogrammed (see 9.1.4). We shall refer to J_{max} as the *maximum degree of multiprogramming*. Thus, whenever J_{max} programs are already being multiprogrammed, additional request for program execution will have to be kept waiting outside the multiprogramming mix; therefore, an additional state for a program is

waiting: waiting to be multiprogrammed.

We assume that programs in this state are kept waiting on M_1.

9.1.4 ***Time Slicing*** A program loses control of the processor either when it is completed or when the total time it has spent in running state since its last loading in M_0 reaches a maximum value Q. If this happens and if there are *waiting* programs, the program will lose all its pages accumulated in M_0, and will join the queue of *waiting* programs until it is allocated an additional quantum Q. If there is no *waiting* program, the program keeps its pages in M_0.

9.1.5 ***Eschenbach Drum Scheme*** (Weingarten, 1966) The tracks of M_1 may be written and read in parallel by fixed heads. We assume that each track is divided into S sectors, each sector containing exactly one page. Each request for a page transfer from M_1 to M_0 is put into a queue associated with the sector containing the requested page. Requests for page transfer are not served in the exact order of arrival; the request with the highest priority for service is the request at the head of the queue associated with the sector that is nearest the head position. This strategy, now widely used, maximizes the number of requests that can be served during each rotation of the memory.

9.2 User Program Model

Let us define n_1 as the current number of programs in the system, i.e., as the total number of programs in one of the four states *waiting*, *ready*, *running*, and *suspended.* With the model set up in the preceding section, the behavior of the user programs can be represented by the state transition diagram shown in Fig. 9.2.

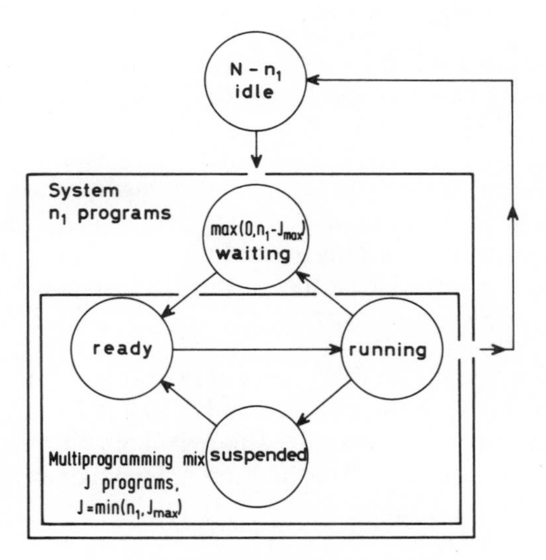

FIGURE 9.2 State transition diagram.

We have assumed that there is a fixed number N of programs cycling in this diagram; the $N-n_1$ programs not present in the system will be considered, for the sake of convenience, as being in an *idle* state.

The behavior of the user programs is supposed to be stochastically defined by the following random variables:

(1) The *user reaction time*, i.e., the idle period or the time interval elapsing between the completion of a user program execution requested by a terminal and the generation of the next request by the same terminal.

(2) The *program execution time*, i.e., the total amount of processor time consumed by a user program between two successive idle periods.

(3) For $n = 1, 2, \ldots$, the amount of processor time, say ξ_n, consumed by a program between the first reference to the nth new page not already referenced, and the $(n+1)$th new page not already referenced. This representation of the *paging activity* of user programs has been chosen because it makes possible, as we shall see later, the evaluation of the effect that time slicing has on the page fault rate.

The distributions of these random variables are determined in the next section.

9.3 Simplifying Assumptions

Our model takes liberties with reality on a certain number of points:

9.3.1 Pages are assumed to be uniformly distributed over the S sectors of M_1; the probability of a demanded page being located in a specific sector is thereby equal to S^{-1}.

9.3.2 We suppose that the traffic of requested pages from M_1 to M_0 is not hindered by the opposite traffic of pages from M_0 to M_1. This assumption is valid if a page frame is permanently maintained vacant in M_0 as well as in each sector of M_1 (Smith, 1967). Under these conditions, we only need to take into account the traffic of pages from M_1 to M_0.

9.3.3 The maximum number of distinct pages a program may accumulate in M_0 at some time t is a function of the number J, $1 \leqslant J \leqslant J_{\max}$, of programs being multiprogrammed at that time t. This maximum number is supposed to be the same for each program of the multiprogramming mix and equal to

$$\mathrm{K}(J) = \mathit{integer}(c_0/J).$$

9.3.4 The *program execution times* are supposed to be independent identically distributed random variables (iidrv), with negative exponential distribution of parameter ϕ.

9.3.5 We define an *execution interval* as the length of a time interval spent by a program in *running* state. An execution interval is at most equal to Q (see 9.1.4).

As a result of assumption 9.3.4, an execution interval x is an iidrv with distribution function, say $B(t)$, equal to

$$B(t) = \Pr(x \leqslant t) = \begin{cases} 0, & t \leqslant 0, \\ 1 - e^{-\phi t}, & 0 < t < Q, \\ 1, & Q \leqslant t. \end{cases}$$

Let μ^{-1} denote the mean length of an execution interval:

$$\begin{aligned}\mu^{-1} &= \int_0^\infty t\, dB(t) \\ &= \int_0^Q t\phi e^{-t}\, dt + Qe^{-\phi Q} \\ &= (1-e^{-\phi Q})\phi^{-1}. \qquad (9.1)\end{aligned}$$

$B(t)$ will be approximated by an exponential distribution of parameter μ defined by (9.1) as a function of ϕ and Q.

9.3.6 User reaction times are considered as iidrv exponentially distributed with parameter λ; the validity of this assumption has been discussed by Coffman and Wood (1966).

9.3.7 Random variables ξ_n, $n = 1, 2, \ldots$, are considered as exponentially distributed iidrv. The parameters of the distributions will be denoted θ_n, $n = 1, 2, \ldots$, respectively.

9.4 Numerical Data

9.4.1 Our model is now defined in terms of the parameters c_0, S, T_{rot}, Q, and $J_{\max}$. Although our approach will in principle be adequate for investigating the influence of any of these parameters on the system performances, we shall mainly center our analysis around the part played by $J_{\max}$ and Q. In order to be able to put our results in concrete form, we shall attribute to each of the other parameters a fixed value, namely,

$$c_0 = 48 \quad \text{page frames}, \qquad S = 4 \quad \text{sectors}, \qquad T_{\text{rot}} = 20 \times 10^{-3} \quad \text{sec.}$$

Likewise, as far as the load on the system is concerned, we shall restrict ourselves to the study of the system sensitivity to fluctuations of the number N of active user terminals. The parameters of the distribution functions that define the stochastic behavior of users and their programs are assigned values inferred from statistical observations. At the time this example was elaborated, available statistics were those observed at SDC (Fine *et al.*, 1966; Totschek, 1965), on the Q-32 time-sharing system (Schwartz *et al.*, 1964; Schwartz and Weissman, 1967):

(1) The average program execution time ϕ^{-1} consumed per user program is taken equal to 1.39 sec.

(2) The average user reaction time λ^{-1} is taken equal to 32 sec.

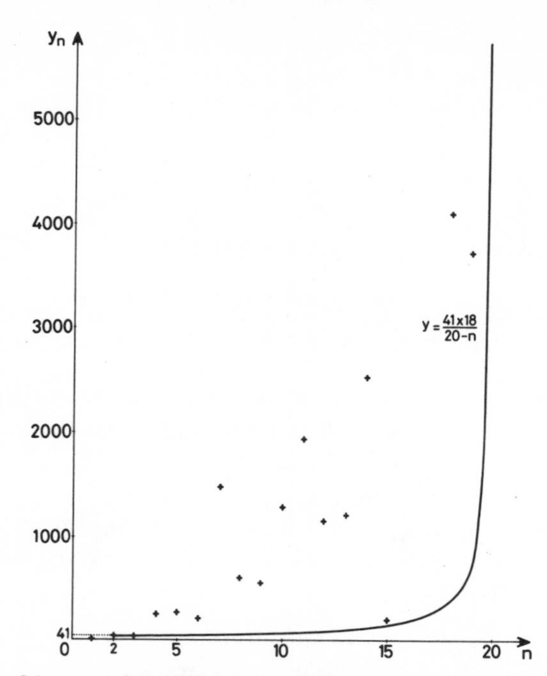

FIGURE 9.3 Mean number of instructions executed between two successive references to a page not already referenced, as a function of the number of distinct pages already referenced.

(3) The paging activity of user programs in the course of their execution is given in Fig. 9.3, which has been obtained from the data reported in Fine *et al.* (1966). It is based on a page size of 1024 words. The mean number y_n of instructions executed between two successive references to an nth and an $(n+1)$th new page not already referenced is plotted against n. The exponential parameters θ_n are given the values

$$\theta_n = (y_n \times 5.10^{-6})^{-1}, \qquad n = 1, 2, \ldots, 19,$$

where 5×10^{-6} sec is the average total execution time of an instruction in the Q-32 time-sharing system. No more than 20 distinct pages are supposed to be referenced during a program execution, so that $\theta_{20} = 0$.

9.4.2 ***Remark*** The branch of hyperbola of Fig. 9.3 is representative of programs that uniformly distribute their references over their pages; more precisely, once their two first pages are referenced they switch uniformly, on the average every 41 references, to any of the 18 other pages. The probability of these programs making a reference to a page already referenced increases with n, the number of distinct pages already referenced, much more slowly than for the programs actually observed. This gives evidence that these latter programs enjoy the *property of locality* already discussed in Section 7.5; at each instant of their execution, they favor a subset of their pages. The values assigned to θ_n, $n = 1, 2, \ldots$, are therefore representative of this locality behavior.

9.5 Page Fault Rate

The page fault rate is the number of pages being requested from M_1 per time unit by a program in the running state. In the evaluation of this rate two cases must be distinguished depending on whether the current number n_1 of programs in the system is larger than J_{max} or not.

If $n_1 > J_{max}$, the running program loses all its pages in M_0 when its execution interval has run out. The number of page faults occurring during an execution interval is therefore at least equal to the number of distinct pages referred to during this interval. If more than $K(J_{max})$ distinct pages are referenced during an execution interval, pages must superimpose each other, and the page fault number might then exceed the number of distinct pages referenced. Following the approach taken by Smith (1967), the average number of page faults per execution interval, say $f(\mu, J_{max})$, can be taken equal to one (the loading of the initial page) plus the number of events of a nonhomogeneous Poisson process with parameters $\theta_1, \theta_2, \ldots, \theta_{K(J max)}$ occurring within a time interval exponentially distributed with parameter μ:

$$f(\mu, J_{max}) = 1 + \frac{\theta_1}{\theta_1 + \mu} + \cdots + \prod_{n=1}^{K(J_{max})-1} \frac{\theta_n}{\theta_n + \mu}$$

$$+ \left\{1 - \frac{\theta_{K(J_{max})}}{\theta_{K(J_{max})+\mu}}\right\}^{-1} \prod_{n=1}^{K(J_{max})} \frac{\theta_n}{\theta_n + \mu},$$

or

$$f(\mu, J_{max}) = 1 + \sum_{z=1}^{K(J_{max})-1} \prod_{n=1}^{z} \frac{\theta_n}{\theta_n + \mu} + \frac{\theta_{K(J_{max})}}{\mu} \prod_{n=1}^{K(J_{max})-1} \frac{\theta_n}{\theta_n + \mu}. \tag{9.2}$$

It is assumed in these expressions that, after a first page superimposition occurs, i.e., after the occurrence of the $K(J_{max})$th page fault, the page fault rate remains constant and the Poisson process is homogeneous with parameter $\theta_{K(J_{max})}$.

Moreover, these expressions imply that the page fault number is the same whatever page replacement rule is used. This is the price paid to be able to take into account the influence of the execution interval on the page fault rate.

Were we interested in the influence of the replacement rule, we should have taken another more appropriate approach to evaluate this page fault rate—a possibility that, of course, is not at all excluded by our global model.

Thus, in this first alternative, when $n_1 > J_{max}$, the page fault rate, say $F(\mu, J_{max})$, is given by

$$F(\mu, J_{max}) = \mu \times f(\mu, J_{max}). \tag{9.3}$$

In the second alternative, when $n_1 = J \leqslant J_{max}$, the J programs that are multiprogrammed are not deprived of the pages they have accumulated in M_0 at the end of each execution interval. The page fault rate is then given by

$$F(\phi, J) = \phi \times f(\phi, J). \tag{9.4}$$

The rates $F(\mu, J_{max})$ and $F(\phi, J)$ are of course averages, which make sense only if the system spends enough consecutive time in the state $n_1 > J_{max}$ or in the state $n_1 \leqslant J_{max}$. We shall see later that this condition is verified precisely because of the near-complete decomposability of the model.

9.6 Parachor and Parachron

Some basic properties of the page fault rate may be found in expression (9.2) where the coefficients θ_n, $n = 1, 2, \ldots$, are representative of the locality property, this general property of program-paging activity. Considering only the $n_1 > J_{max}$ case and denoting by $\Delta F(\mu, J_{max})$ the page fault rate difference that would result from an allotment of an additional page frame ($K(J_{max}) := K(J_{max}) + 1$) to each multiprogrammed user program, from (9.3) we obtain

$$\Delta F(\mu, J_{max}) = \mu \times \Delta f(\mu, J_{max}),$$

and from (9.2) we have

$$\Delta f(\mu, J_{max}) = \prod_{n=1}^{K(J_{max})} \frac{\theta_n}{\theta_n + \mu} + \frac{\theta_{K(J_{max})+1}}{\mu} \prod_{n=1}^{K(J_{max})} \frac{\theta_n}{\theta_n + \mu}$$
$$- \frac{\theta_{K(J_{max})}}{\mu} \prod_{n=1}^{K(J_{max})-1} \frac{\theta_n}{\theta_n + \mu}.$$

After a few algebraic simplifications, these expressions reduce to

$$\Delta F(\mu, J_{max}) = (\theta_{K(J_{max})+1} - \theta_{K(J_{max})}) \prod_{n=1}^{K(J_{max})} \frac{\theta_n}{\theta_n + \mu}. \tag{9.5}$$

A first well-known consequence we can draw from (9.5) is that the *page fault rate is a decreasing function of the primary memory space guaranteed to each multiprogrammed program*. Indeed, the shape of the paging activity displayed in Fig. 9.3 is, on the average, such that

$$y_1 \leqslant y_2 \leqslant \cdots \leqslant y_{K(J_{\max})}, \qquad 1 \leqslant K(J_{\max}) \leqslant 20.$$

Hence

$$\theta_1 \geqslant \theta_2 \geqslant \cdots \geqslant \theta_{K(J_{\max})}, \tag{9.6}$$

and $\Delta F(\mu, J_{\max}) \leqslant 0$.

Moreover, $\Delta F(\mu, J_{\max}) \to 0$ as $K(J_{\max})$ increases: since $\theta_n/(\theta_n+\mu) < 1$, the product

$$\prod_{n=1}^{K(J_{\max})} \frac{\theta_n}{\theta_n+\mu} \tag{9.7}$$

decreases as $K(J_{\max})$ increases; and, because of inequalities (9.6), this decrease is even faster than the power $K(J_{\max})$.

The fact that expression (9.2) takes into account the *influence of the execution interval upon the page fault rate* is more interesting. The product (9.7) and thus the difference $\Delta F(\mu, J_{\max})$ decrease as $\mu \to \infty$; the larger $K(J_{\max})$ is, the faster this decrease is: for large values of μ, (9.7) behaves approximately as $\mu^{-K(J_{\max})}$.

On the other hand, a result of (9.2) is that

$$\lim_{\mu \to 0} F(\mu, J_{\max}) = \lim_{\mu \to 0} \mu \times f(\mu, J_{\max}) = \theta_{K(J_{\max})},$$

and

$$\lim_{\mu \to \infty} f(\mu, J_{\max}) = 1,$$

so that, for large values of μ, $F(\mu, J_{\max}) \approx \mu$.

Because of inequalities (9.6), $\theta_{K(J\max)}$ is a minimum value of θ_n for a fixed value of $K(J_{\max})$.

Therefore, *for any given value $K(J_{\max})$ of the primary memory space guaranteed to each multiprogrammed program, the page fault rate tends to a minimum $\theta_{K(J_{\max})}$ as the average execution interval is increased until it becomes equal to the average program execution time; the page fault rate ends up increasing as the inverse of the average executive interval when this latter approaches zero.*

The variations of the page fault rate $F(\mu, J_{\max})$ as a function of $K(J_{\max})$ and of the average execution interval μ^{-1} are illustrated by Figs. 9.4 and 9.5. Despite their irregularities—which result from the irregularities of y_n—these figures show what should be the *parachor* and what we could call the *parachron* of our user programs. Parachor is a term introduced by Belady and Kuehner (1969) to designate the amount of primary memory needed by a program to maintain a "satisfactory" rate of page faults; similarly we call an execution interval that suffices to obtain a satisfactory page fault rate a *parachron*. These two quantities are of course mutually related. Figures 9.4 and 9.5 show,

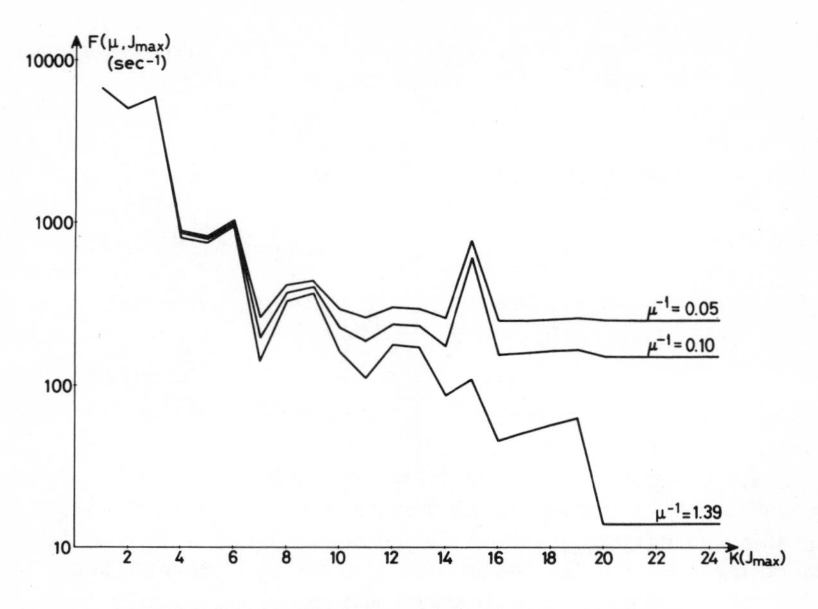

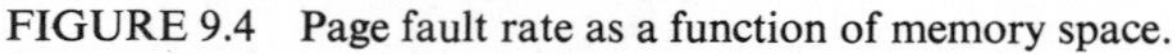
FIGURE 9.4 Page fault rate as a function of memory space.

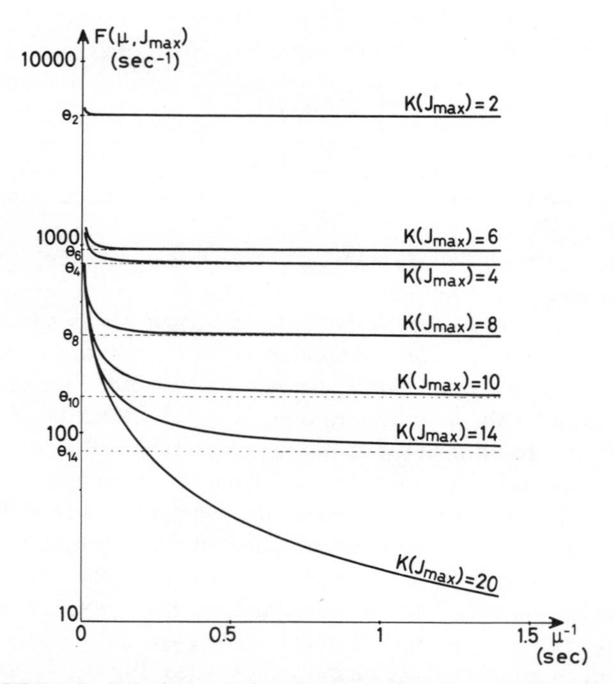

FIGURE 9.5 Page fault rate as a function of average execution interval.

however, that as long as $K(J_{max})$ remains below 10 (i.e., below half the maximum program size) there is not much to be gained by allocating a program a parachron of more than 0.3 to 0.4 sec and a parachor of more than about 7 page frames; conversely, these figures show also that the parachor should at least be of 4 pages and the parachron of 0.05 sec.

The reason we insist on these values is that, as we shall see in Section 9.9.4, the parachor and parachron values, *which are satisfactory measures for programs considered as individuals, are far from optimizing performance criteria that take page transfer rates into account.* What is satisfactory or optimal for programs considered as individuals will not appear so when they are considered collectively.

9.7 Page Transfer Rate

We have seen in Section 9.1.5 that requests for page transfer from M_1 to M_0 are not served in the exact order of their arrival. In order to maximize the number of requests served per rotation, the request with highest priority for service is the request at the top of the queue associated with the sector nearest the current head position (Eschenbach drum scheme). The result of this strategy is that the number of requests served per rotation, i.e., the page transfer rate, is a function of the number of pending requests. This rate is obviously null if no request is pending and approaches asymtotically S/T_{rot} (as many requests served per rotation as there are sectors) when the number of pending requests increases beyond limit. Because of simplifying assumptions

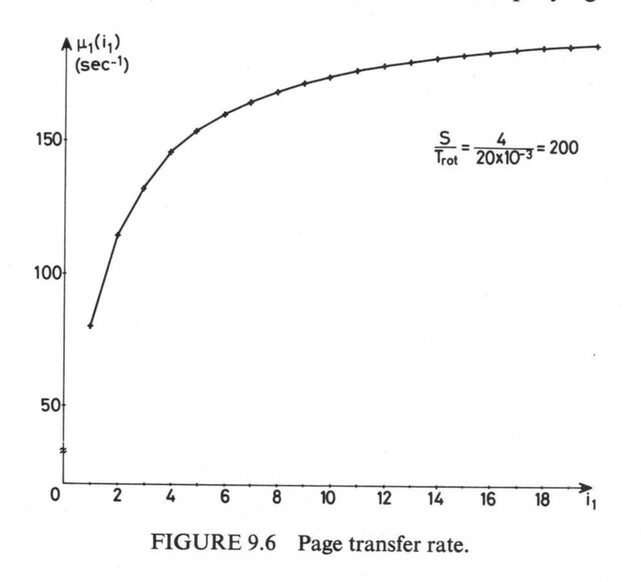

FIGURE 9.6 Page transfer rate.

9.3.1 and 9.3.2 this function can be easily calculated. It was first calculated by Smith (1967).[1] In our model the number of pending requests is equal to the number of programs in *suspended* state. Let us use i_1 to denote the number of programs in suspended state $(0 \leqslant i_1 \leqslant J)$, and $\mu_1(i_1)$ to denote the page transfer rate. We have (see the appendix in Smith (1967) for a proof of this result):

$$\mu_1(i_1) = \frac{2i_1}{S+2i_1-1} \times \frac{S}{T_{\text{rot}}}, \qquad 0 \leqslant i_1 \leqslant J. \tag{9.8}$$

The shape taken by $\mu_1(i_1)$ as a function of i_1 is shown in Fig. 9.6 for the particular values given in our model to S and T_{rot}.

The point we want to make here is to show how, by means of state-dependent service rates, our model can take into account strategies that tend to minimize access delays to secondary rotational memories. Expressions similar to (9.8) could for example be deduced from Denning (1967) for movable-head disks.

9.8 Decomposability of the Model

The rates of transition between *running*, *suspended*, and *ready* states will be, for most systems that correspond to our model, much higher than the rates at which programs are created and completed, i.e., than the rates at which programs leave and return to the *idle* state. We may therefore suspect that our model will in general be nearly completely decomposable into the set of N terminals on the one hand and an *aggregate* corresponding to the subsystem CPU–primary and secondary memory on the other hand. Let us verify under which conditions this is true for our model.

As a result of the simplifying assumptions we made in Section 9.3 our hypothetical system may be regarded as a closed network of queues in which a fixed number N of programs cycle between three stages of service. If i_l, $l = 0, 1, 2$, is used to denote the number of programs at stage l, we have

i_0 is the number of programs in *ready* or *running* state;
i_1 is the number of programs in *suspended* state;
i_2 is the number of programs in *idle* state.

Summing up our previous definitions, we have

$$i_2 = N - n_1 \qquad \text{and} \qquad i_0 + i_1 = J = \min(n_1, J_{\max}),$$

$n_1 - J_{\max}$ being the number of programs in *waiting* state. Keeping the same

[1] Smith's results were later rediscovered by, among others, Burge and Konheim (1971) and Denning (1972).

notation as in Chapter IV, we use $\mu_l p_{lm}$, $l, m = 0, 1, 2$, to denote the transfer rate from stage l to stage m, $l \neq m$; thus $\mu_l p_{lm}$ is the probability that a program leaves state l for state m during any single time unit chosen sufficiently small. Taking the microsecond, we have

$$\begin{aligned}
\mu_0 p_{01} &= 10^{-6} F(\mu, J_{\max}) \quad \text{if} \quad n_1 > J_{\max}, \\
&= 10^{-6} F(\phi, J) \quad \text{if} \quad n_1 = J \leqslant J_{\max} && \text{(Section 9.5)}, \\
\mu_0 p_{02} &= 10^{-6} \phi && \text{(assumption 9.3.4)}, \\
\mu_1 p_{10} &= 10^{-6} \mu_1(i_1), \quad 0 \leqslant i_1 \leqslant J && \text{(expression (9.8))}, \\
\mu_2 p_{20} &= 10^{-6} \lambda i_2, \quad 0 \leqslant i_2 \leqslant N && \text{(assumption 9.3.6)}.
\end{aligned}$$

Our criterion of near-complete decomposability stipulates that the maximum degree of coupling of the aggregate with the set of terminals must be less than half the minimum indecomposability of the aggregate. The maximum degree of coupling is given by

$$\max(\mu_0 p_{02}, \mu_2 p_{20}) = 10^{-6} \max(\phi, \lambda i_2) = 10^{-6} \max[\phi, (N-1)\lambda],$$

and the aggregate indecomposability is given by (see Section 4.3)

$$\mu_0 p_{01} + \mu_1 p_{10} - 2(\mu_0 p_{01} \mu_1 p_{10})^{1/2} \cos \frac{\pi}{n_1 + 1}. \tag{9.9}$$

In (9.9) the page fault rate $\mu_0 p_{01}$ increases faster than the cosine term when $n_1 = J \leqslant J_{\max}$ (the number of multiprogrammed jobs) increases. The minimum is thus attained for $n_1 = 1$ and is given by (see Figs. 9.4 and 9.6)

$$10^{-6}[F(\phi, 1) + \mu_1(1)] \approx 10^{-6}(14 + 80).$$

Hence, our condition for near-complete decomposability reduces to

$$\max[\phi, (N-1)\lambda] \leqslant \tfrac{1}{2}(14 + 80); \tag{9.10}$$

and if $N - 1 > \phi/\lambda = 32/1.39 = 23$, this condition becomes

$$(N-1)\lambda \leqslant 47,$$

which remains verified as long as $N \leqslant 1504$.

This shows how a prior knowledge of the values of a few parameters characterizing the hardware and the work load enables the near-complete decomposability of a given system to be verified. Moreover, the approximation made when using an aggregative model can also be estimated. This approximation is here of the order of

$$\frac{\max[\phi, (N-1)\lambda]}{F(\phi, 1) + \mu_1(1)} \approx 10^{-3} \max\left(7, \frac{N-1}{3}\right).$$

9.9 Aggregate Short-Term Equilibrium

When condition (9.10) is verified, our system is nearly completely decomposable into the set of user terminals on the one hand, and a CPU–primary and secondary memory aggregate on the other hand. The short-term equilibrium attained by the aggregate may then be studied independently of the interactions with the user terminals; the long-term equilibrium attained ultimately by these interactions will be studied later in terms of the equilibrium properties of the aggregate.

Since in this short-term equilibrium analysis the activity of the user terminals can be disregarded, we can assume that the number of programs in the aggregate keeps a fixed and constant value n_1, $1 \leqslant n_1 \leqslant N$. Following the notation used in Chapter V, we shall use $\mathscr{M}_1(n_1)$ to refer to this aggregate.

It was shown in Section 9.5 that depending on whether or not $n_1 > J_{\max}$, the page fault rate was different. For this reason we have to consider two distinct cases:

(1) $n_1 \leqslant J_{\max}$, i.e., the aggregate $\mathscr{M}_1(n_1)$ is underloaded. No program is *waiting* and the $n_1 = J$ programs are multiprogrammed without execution interval; the page fault rate is equal to $F(\phi, J)$.

(2) $n_1 > J_{\max}$, i.e., $\mathscr{M}_1(n_1)$ is overloaded. $n_1 - J_{\max}$ programs are *waiting* and $J_{\max}$ programs are multiprogrammed, losing their pages in M_0 at the end of each execution interval. The page fault rate is equal to $F(\mu, J_{\max})$.

Considering the underload case first, the short-term equilibrium distribution of programs among *ready* or *running* and *suspended* states in $\mathscr{M}_1(n_1)$ can be obtained as the steady-state distribution of the congestion in an $M \mid M \mid 1 \mid J$ queueing system with fixed and finite population $J = n_1$, service rate $F(\phi, J)$, and congestion-dependent input rate $\mu_1(i_1)$, $i_1 = 0, \ldots, J$, and this for all values $n_1 = J$, $1 \leqslant J \leqslant J_{\max}$. Since the page transfer rate $\mu_1(i_1)$ is dependent on the congestion, this steady-state distribution is defined by relations (6.6). Introducing into (6.6) the rates $F(\phi, J)$ and $\mu_1(i_1)$, for $J = 1, \ldots, J_{\max}$ yields

$$\pi_1(i_0 \mid J) = \frac{\prod_{i_1=J-i_0+1}^{J} \mu_1(i_1)}{[F(\phi, J)]^{i_0}} \pi_1(0 \mid J), \qquad i_0 = 1, \ldots, J, \tag{9.11}$$

with

$$\pi_1(0 \mid J) = \left\{1 + \sum_{j=1}^{J} \frac{\prod_{i_1=J-j+1}^{J} \mu_1(i_1)}{[F(\phi, J)]^j}\right\}^{-1},$$

where $\pi_1(i_0 \mid J)$ is the probability of i_0 programs in *ready* or *running* state given that J programs are multiprogrammed.

In particular,

$$\sigma(J) = 1 - \pi_1(0 \mid J), \qquad J = 1, \ldots, J_{\max}, \tag{9.12}$$

is the probability of the processor not being idle when J programs are multiprogrammed. This probability will be henceforth referred to as the *processor efficiency*.

Similarly, for the overload case when $n_1 > J_{max}$, we obtain a distribution that we denote

$$\{\pi_1{}^*(i_0 \mid J_{max})\}, \qquad i_0 = 0, \ldots, J_{max},$$

by replacing J by J_{max} and $F(\phi, J)$ by $F(\mu, J_{max})$ in expressions (9.11). The processor efficiency will in this case be denoted

$$\sigma^*(J_{max}) = 1 - \pi_1{}^*(0 \mid J_{max}). \tag{9.13}$$

In Fig. 9.7, $\sigma^*(J_{max})$ is displayed for values of J_{max} up to 20 and for four distinct values of the average execution interval μ^{-1}. When $\mu^{-1} = \phi^{-1} = 1.39$ sec, $\sigma^*(J_{max})$ is identical to $\sigma(J)$; otherwise when $\mu^{-1} < \phi^{-1}$, for any J_{max} we have

$$\sigma^*(J_{max}) < \sigma(J_{max})$$

since the page fault rate $F(\mu, J_{max})$ increases with μ (see Section 9.6).

9.9.1 ***Optimal Degrees of Multiprogramming*** We see from Fig. 9.7 that for almost each value of the average execution interval, there is a different J_{max} that maximizes $\sigma^*(J_{max})$. We shall use J_{max}^{opt} to denote the value of J_{max} that, for a given μ, maximizes $\sigma^*(J_{max})$ or, more precisely, maximizes the sum

$$\sum_{i_0=1}^{J_{max}} \frac{\prod_{i_1=J_{max}-i_0+1}^{J_{max}} \mu_1(i_1)}{[F(\mu, J_{max})]^{i_0}},$$

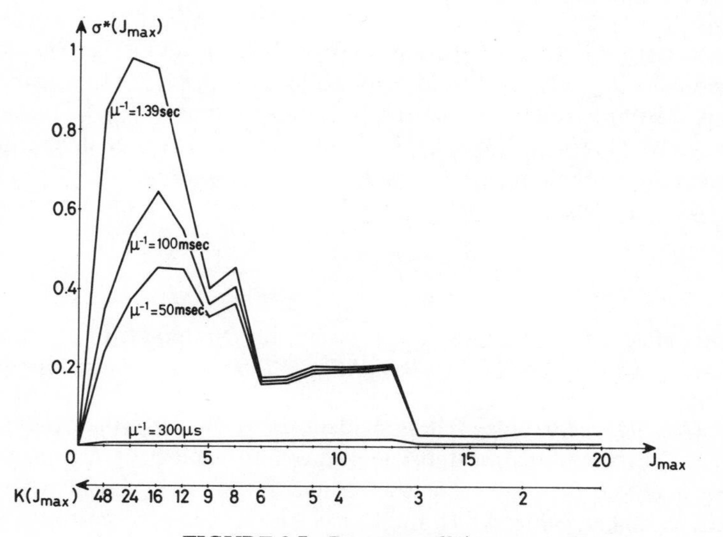

FIGURE 9.7 Processor efficiency.

and we shall use J_0 to denote the J value that maximizes $\sigma(J)$, that is, that maximizes the sum

$$\sum_{i_0=1}^{J} \frac{\prod_{i_1=J-i_0+1}^{J} \mu_1(i_1)}{[F(\phi, J)]^{i_0}}.$$

An essential property of near-complete decomposability is that the relative values of the short-term equilibrium attained in an aggregate are approximately maintained in the long run. Therefore, these optimal values are also those which maximize in the long run the fraction of time the processor is busy, with full or incomplete load.

The existence of these optimal values depends on the relative importance of two well-known antagonistic effects that take place when the number of multiprogrammed jobs increases: the decrease of the probability of the processor remaining idle on the one hand, and the increase of the page fault rate as the space available to each program in main memory shrinks on the other hand. The advantage of our approach here is that these optimal values may be obtained without neglecting the influences of the locality property and of the additional paging activity caused by the enforcement of a time slice.

Figure 9.7 indicates that the *optimal maximum degree of multiprogramming tends to increase as the average execution interval* μ^{-1} *decreases.* This is explained by the fact that the page fault rate does not start to increase before programs have exhausted their primary memory space $K(J_{max})$. The shorter the average execution interval, the smaller this exhaustible space is; a smaller $K(J_{max})$ corresponds to a larger J_{max}, of course.

9.9.2 ***Remark*** Figure 9.7 also shows that $\sigma^*(J_{max})$ [and $\sigma(J_{max})$] is not only increasing for $J_{max} \leqslant J_{max}^{opt}$, but also within intervals of J_{max} for which $K(J_{max})$ remains constant, and thus also the page fault rate $F(\mu, J_{max})$. By means of expression (9.11) for $\pi_1{}^*(0 \mid J_{max})$, it is indeed easy to prove, since $\mu_1(i_1)$ is a nondecreasing function of i_1, that $\sigma^*(J_{max}) < \sigma^*(J_{max}+1)$ if $K(J_{max}) = K(J_{max}+1)$, i.e., when

$$\text{integer}\left(\frac{c_0}{J_{max}}\right) = \text{integer}\left(\frac{c_0}{J_{max}+1}\right).$$

The only effect of an increase of J_{max} within these intervals of constant page fault rate is the decrease of the probability of the processor remaining idle.

9.9.3 ***Overload: Thrashing*** It is evident from Fig. 9.7 that the primary memory allotment that must be guaranteed to each program in order to achieve maximal processor efficiency under full load is considerably larger than the parachor value, which in Section 9.6 has been shown to be between 4 and 7 pages.

The difference is justified by the fact that the parachor is an individual program characteristic, whereas $\sigma(J)$ and $\sigma^*(J_{max})$ are measures of the overlapping achieved between page transfers and processor busy periods, and depend on secondary memory transfer rates as well.

Denning's (1968b) argument on thrashing helps us to understand why $K(J_{max}^{opt})$ is larger than the parachor. A measure e of the ability of a program to use the processor may be defined as

$$e = \frac{\mu^{-1}}{\mu^{-1} + f(\mu, J_{max}) \times \mu_1^{-1}}, \tag{9.14}$$

where μ^{-1} is the average execution interval, $f(\mu, J_{max})$ the average number of page faults per execution interval, and μ_1^{-1} is used to denote the average time needed by a page transfer:

$$\mu_1 = \sum_{i_0=0}^{J_{max}-1} \pi_1^*(i_0 \mid J_{max})\, \mu_1(J_{max} - i_0).$$

Thus, $f(\mu, J_{max}) \times \mu_1^{-1}$ is the average time a program spends waiting for page transfers during an execution interval.

If we multiply (9.14) above and below by μ, we have

$$e = \frac{1}{1 + F(\mu, J_{max}) \times \mu_1^{-1}},$$

and the slope of e in function of the page fault rate is

$$\frac{\Delta e}{\Delta F(\mu, J_{max})} = -\frac{\mu_1^{-1}}{[1 + F(\mu, J_{max})\mu_1^{-1}]^2}. \tag{9.15}$$

This shows that, as long as $F(\mu, J_{max})\mu_1^{-1}$ is not much larger than one, e is all the more sensitive to variations of the page fault for μ_1^{-1} being larger. The longer the page transfer time is, the smaller the page fault rate must be to achieve a given level of processor efficiency. Large μ_1^{-1} values are responsible for values of the primary memory space $K(J_{max}^{opt})$ exceeding the parachor.

This extreme sensitivity of processor efficiency to variations of the page fault rate when the page transfer rate is low may cause under full load serious degradations of the system response time when the degree of multiprogramming is allowed to exceed J_{max}^{opt}. These degradations will be more carefully analyzed in the following sections.

9.9.4 ***Overload: Influence of the Execution Interval*** Figure 9.7 shows that in the overload case *the longer the mean execution interval, the smaller is the primary memory space* $K(J_{max})$ *required to attain a specified processor efficiency.* Conversely, the larger $K(J_{max})$, the more advantageous it is from the point of view of processor efficiency to provide for the longest possible execution intervals. This is a consequence of the increasingly preponderant influence μ has on the page fault rate decrement $\Delta F(\mu, J_{max})$ as $K(J_{max})$ increases [see Eq. (9.5) and inequalities (9.6)].

It can be more intuitively understood by the fact that when $K(J_{max})$ is sufficiently large, the locality property of programs is fully exploited and a limited execution interval then becomes the main source of page faults. One can, for example, observe in Fig. 9.7 that $\sigma^*(J_{max})$ is practically insensitive to J_{max} if μ^{-1} is so small (in this example less than about $\sum_{i=1}^{3} \theta_i^{-1} \approx 385$ μsec) that programs are not given enough time, whatever $K(J_{max})$ is, to accumulate in M_0 the minimum number of pages necessary for the property of locality to have effective results (about 4 pages, as shown in Fig. 9.3).

Therefore, a policy that optimizes processor efficiency by taking advantage of the locality property should provide the user program with the longest possible execution intervals. It is remarkable that such a policy will give the most effective results when the primary memory allotment per program is large, viz., when the degree of multiprogramming is small. A generous main-memory allocation demands an equally generous processor time allocation. This is indeed an additional argument in favor of the processor allocation mechanisms that have already been advocated by Mullery and Driscoll (1970). These mechanisms minimize the number of changes of tasks (and thereby reduce overheads on processors) when the system is underloaded, viz., when the number of outstanding requests for program execution is less than the number of quanta contained within a prerequisite response time.

9.10 System Long-Term Equilibrium

Let us now study the long-term dynamics of the interactions between the user terminals and the aggregates $\mathcal{M}_1(n_1)$, $n_1 = 0, \ldots, N$. Because our system is supposed to be nearly completely decomposable, the local equilibrium states of the aggregates, attained in the short run, are approximately preserved in the long run. The entire system can thus be regarded as a set of N active user terminals originating requests of program execution for the aggregates $\mathcal{M}_1(n_1)$, each aggregate being in a state of equilibrium defined by the probability distribution $\pi_1(i_0 \mid n_1)$ if $n_1 \leqslant J_{max}$, or $\pi_1{}^*(i_0 \mid J_{max})$ if $n_1 > J_{max}$ (cf. Section 9.9). Here n_1 is the number of programs in the system being or waiting to be multiprogrammed; we will henceforth refer to n_1 as the *system congestion*.

Let us use $\psi(n_1)$ to denote the rate at which program executions are completed in the aggregate $\mathcal{M}_1(n_1)$; it follows from (9.12) and (9.13) that this rate is given by

$$\psi(n_1) = \begin{cases} \phi[1-\pi_1(0 \mid n_1)] = \phi \times \sigma(n_1) & \text{if } n_1 \leqslant J_{max} \quad \text{(underload)}, \\ \phi[1-\pi_1{}^*(0 \mid J_{max})] = \phi \times \sigma^*(J_{max}) & \text{if } n_1 > J_{max} \quad \text{(overload)}, \end{cases} \tag{9.16}$$

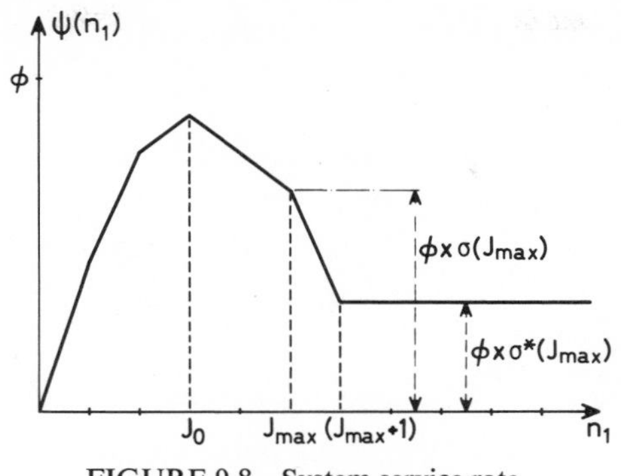

FIGURE 9.8 System service rate.

where $\sigma(n_1)$, $n_1 \leqslant J_{\max}$, and $\sigma^*(J_{\max})$ are the processor efficiency for the underload and the overload case, respectively, and ϕ^{-1} is the average total processor time consumed by a program.

We will refer to $\psi(n_1)$, which in fact is here the equivalent of the probability $\psi_{1,2}(n_1)$ introduced in Eq. (5.3), as the *system service rate*. If we neglect the slight variations of $\sigma^*(J_{\max})$ discussed in Remark 9.9.2, the more general shape of $\psi(n_1)$ is given by Fig. 9.8; $\psi(n_1)$ increases with n_1 up to the congestion J_0 that maximizes $\sigma(n_1)$, and then decreases until the congestion $J_{\max}+1$, from and above which it remains constant.

It is important to note that

$$\sigma^*(J_{\max}) \leqslant \sigma(J_{\max}),$$

and thus

$$\psi(J_{\max}+1) \leqslant \psi(J_{\max}).$$

The equality is verified only if no time slicing is enforced, viz., if the quantum Q is infinite ($\mu^{-1} = \phi^{-1}$). Otherwise, the decrease of the service rate, once the system is overloaded ($n_1 \geqslant J_{\max}+1$), measures the consequence of the supplement of page traffic necessary, at the end of each execution interval, to exchange in main memory the pages of the *running* program with those of a *waiting* program.

Hence, $\psi(n_1)$ will be a nondecreasing function of n_1 only if $J_{\max} \leqslant J_0$ *and* no time slicing is applied.

The service rate $\psi(n_1)$ is plotted in Fig. 9.9 for an average execution interval $\mu^{-1} = 0.1$ sec, and for $J_{\max} = 1, 2, \ldots, 5$.

The rate at which requests for program execution are originated from the terminals is equal to

$$(N-n_1)\lambda, \qquad n_1 = 0, \ldots, N, \tag{9.17}$$

where λ^{-1} is the average user reaction time (see 9.3.6). This rate is thus also function of the congestion n_1; it is the *input rate* at which programs enter the aggregate $\mathscr{M}_1(n_1)$. This input rate is shown in Fig. 9.10.

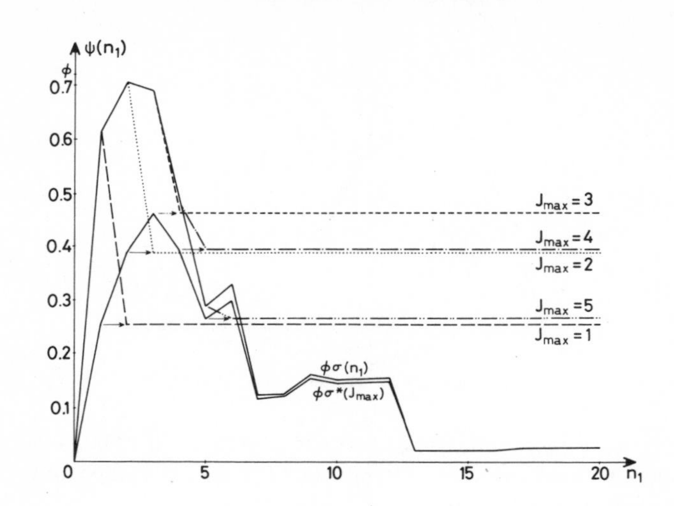

FIGURE 9.9 System service rate for different values of J_{max}; $\mu^{-1} = 100$ msec.

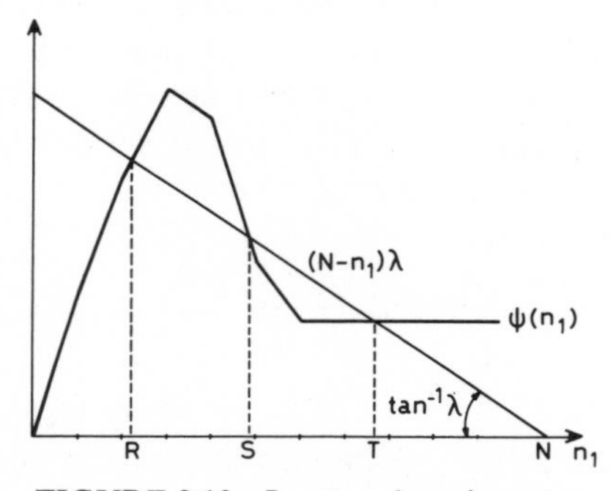

FIGURE 9.10 Input and service rates.

Consequently, a particular property of our aggregate $\mathcal{M}_1(n_1)$ is that both its input rate and its output rate (the service rate) are a function of the congestion n_1. The long-term probability distribution of this congestion n_1, given that there are N active user terminals, can be calculated as the steady-state distribution of the congestion in an $M|M|1|N$ queueing system with finite population N, congestion-dependent service rate $\psi(n_1)$, and input rate $(N-n_1)\lambda$. In agreement with the notation used in Chapter V, let us use $\{\pi_2(n_1|N)\}$, $n_1 = 0, \ldots, N$, to denote this distribution, $\pi_2(n_1|N)$ being the steady-state probability of n_1 jobs in the system, being or waiting to be multiprogrammed. If we introduce (9.16) and (9.17) into (6.6), we obtain

$$\pi_2(n_1|N) = \lambda^{n_1} \frac{N(N-1)\cdots(N-n_1+1)}{\prod_{k=1}^{n_1} \psi(k)} \pi_2(0|N), \qquad n_1 = 1, \ldots, N, \qquad (9.18)$$

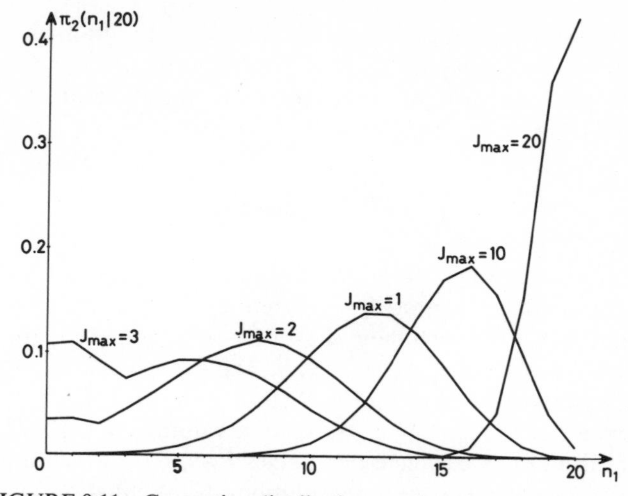

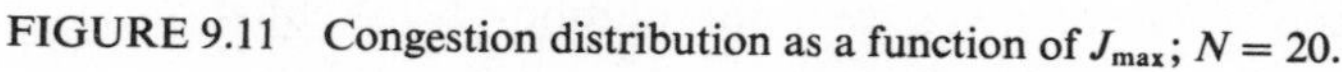
FIGURE 9.11 Congestion distribution as a function of $J_{\max}$; $N = 20$.

with

$$\pi_2(0\,|\,N) = \left[1 + \sum_{j=1}^{N} \lambda^j \frac{N(N-1)\cdots(N-j+1)}{\prod_{k=1}^{j} \psi(k)}\right]^{-1}.$$

This distribution of the congestion is displayed in Fig. 9.11 for $N = 20$ and for some typical values of $J_{\max}$. The distribution $\pi_2(n_1\,|\,N)$ has many interesting properties that will be investigated in the remaining sections of this chapter. Let us just note for now that we can deduce from it the long-term and unconditional distributions for the different states of the aggregate. More precisely, the long-term and unconditional probability of i_0 programs in *ready* or *running* state, when there are N active user terminals, can be obtained as

$$s_0(i_0) = \sum_{n_1=i_0}^{J_{\max}} \pi_2(n_1\,|\,N)\,\pi_1(i_0\,|\,n_1) + \pi_1{}^*(i_0\,|\,J_{\max}) \sum_{n_1=J_{\max}+1}^{N} \pi_2(n_1\,|\,N), \qquad i_0 = 0, \ldots, J_{\max}.$$

The long-term and unconditional probability of i_1 programs in suspended state is

$$s_1(i_1) = \sum_{n_1=i_1}^{J_{\max}} \pi_2(n_1\,|\,N)\,\pi_1(n_1-i_1\,|\,n_1) + \pi_1{}^*(J_{\max}-i_1\,|\,J_{\max}) \sum_{n_1=J_{\max}+1}^{N} \pi_2(n_1\,|\,N), \qquad i_1 = 0, \ldots, J_{\max}.$$

The long-term and unconditional probability of i_1' programs waiting to be multiprogrammed is simply $\pi_2(J_{max}+i_1' \mid N)$, $i_1' = 0, \ldots, N-J_{max}$.

These distributions yield the coefficients of utilization of the processor, $1-s_0(0)$, and of the secondary memory, $1-s_1(0)$.

9.11 Instabilities

As stated above, a peculiarity of the aggregate $\mathcal{M}_1(n_1)$ is that both its input rate $(N-n_1)\lambda$ at which requests for program executions are originated, and its service rate $\psi(n_1)$ at which these executions are completed, are dependent on the system congestion n_1, $n_1 = 0, \ldots, N$. The general shape[1] of these rates is displayed as a function of n_1 on Fig. 9.10.

One observes that, depending on the relative values of N, J_{max}, λ, and $\psi(n_1)$, there may be at most three congestion values R, S, T for which the input rate equates the service rate, viz., that are solutions of the equation

$$(N-n_1)\lambda = \psi(n_1), \tag{9.19}$$

where $\psi(n_1)$ is given by (9.16). Figure 9.10 shows clearly that (9.19) can have either the three solutions R, S,and T, or only one solution, which is then R or T.

It is easy to understand that R and T are values around which the system congestion is inclined to come into equilibrium. Let us consider the system at some instant when the congestion is equal to the integer value nearest R. An increment $(+\Delta n_1)$ of the congestion causes the service rate to exceed the input rate. This excess will tend to reduce the congestion to its original value.

Likewise, a decrement $(-\Delta n_1)$ would cause the input rate to exceed the output rate, compelling the congestion to reincrease. The same reasoning applies to the value T. Thus, the vicinities of R and T are *stochastically stable* domains for the congestion.

Inversely, a similar argumentation indicates that in the vicinity of S the congestion variations are reinforced instead of being deadened by the alterations they cause to the output/input rate ratio. The system will always tend to leave the states in the vicinity of S in favor of congestions near R or T. The vicinity of S is a *stochastically unstable* domain for the congestion. In short, we shall call R and T *stable* solutions of Eq. (9.19), and S an *unstable* solution.

Depending on the relative values of the various system parameters, the unstable solution S may or may not exist. In the first case we can predict that the congestion will become preferentially steady around one stable solution,

[1] Although n_1 is an integer variable, it will often be convenient in the rest of this chapter to consider $\psi(n_1)$ a continuous function of n_1, each interval $[\psi(n_1), \psi(n_1+1)]$ consisting of a segment of a straight line; the same convention will apply to other functions of n_1, N, etc.

R or T. In the second case, when S exists, the two stable solutions R and T coexist and we can predict that the congestion in such systems will experience random fluctuations between the two stable domains around R and T.

This type of behavior may indeed be more precisely formulated. We can deduce from the probability distribution of the congestion defined by (9.18) that for $n_1 = 1, \ldots, N$,

$$\frac{\pi_2(n_1 \mid N)}{\pi_2(n_1 - 1 \mid N)} = \frac{\lambda^{n_1} N(N-1)\cdots(N-n_1+1)\prod_{k=1}^{n_1-1}\psi(k)}{\lambda^{n_1-1} N(N-1)\cdots(N-n_1+2)\prod_{k=1}^{n_1}\psi(k)}$$

or

$$\frac{\pi_2(n_1 \mid N)}{\pi_2(n_1 - 1 \mid N)} = \frac{(N-n_1+1)\lambda}{\psi(n_1)}. \tag{9.20}$$

The probability $\pi_2(n_1 \mid N)$ is thus a nondecreasing function of n_1 as long as $(N-n_1+1)\lambda \geqslant \psi(n_1)$, and a nonincreasing function as long as $(N-n_1+1)\lambda \leqslant \psi(n_1)$. Hence, for a probability, say $\pi_2(n_A \mid N)$, that is a maximum of the distribution, we must have

$$(N-n_A+1)\lambda \geqslant \psi(n_A) \qquad \textit{and} \qquad (N-n_A)\lambda \leqslant \psi(n_A+1), \tag{9.21}$$

and for a minimum of the distribution, say a probability $\pi_2(n_B \mid N)$, we must have

$$(N-n_B+1)\lambda \leqslant \psi(n_B) \qquad \textit{and} \qquad (N-n_B)\lambda \geqslant \psi(n_B+1). \tag{9.22}$$

The extrema of the distribution $\{\pi_2(n_1 \mid N)\}$ correspond therefore to the intersections of the service rate $\psi(n_1)$ and of the straight line $(N-n_1+1)\lambda$. More precisely, each extremum is the largest integer smaller than a solution of

$$(N-n_1+1)\lambda = \psi(n_1). \tag{9.23}$$

This equation also has either three solutions, say A, B, C, with B unstable and A, C stable, or only one stable solution A or C. So, from the inequalities (9.21) and (9.23) we conclude that *the largest congestion n_A or n_C smaller than or equal to a stable solution A or C of* (9.23) *is a congestion of highest probability. Similarly the largest congestion n_B smaller than or equal to the unstable solution B is a congestion of least probability.*

The solutions A, B, C are easily obtained by superimposing on a same diagram, as indicated in Fig. 9.12, the service rate $\psi(n_1)$ and the straight line $(N-n_1+1)\lambda$, which is simply the input rate shifted horizontally by one unit to the right. By comparing Fig. 9.12 with Fig. 9.11, one can verify for $J_{\max} = 3$ and $N = 20$ that the two congestions of highest probability $n_1 = 1$ and $n_1 = 6$, correspond to intersections A and C, respectively, and that the congestion of least probability, i.e., $n_1 = 3$, corresponds to intersection B.

The random fluctuations of the congestion we have predicted above take place around n_B and between the two stable congestions n_A and n_C when the unstable solution B exists, that is, when for fixed values of the other parameters λ, $J_{\max}$, and $\psi(n_1)$ the value $N+1$ lies between the two bounds N^+ and N^{++}, which are graphically determined in Fig. 9.12.

As shown in this figure, the interval $[N^+, N^{++}]$ corresponds to an interval $[n_1^{++}, n_1^{+}]$ of n_1 in which the slope of $\psi(n_1)$ is negative and steeper than $-\lambda$:

$$\Delta\psi(n_1)/\Delta n_1 < -\lambda.$$

This range of N values, i.e.,

$$N^+ - 1 \leqslant N \leqslant N^{++} - 1,$$

is characterized by a relatively large dispersion of the congestion and of the response time of the system. In our model, for example, when $J_{\max} = 3$, these bounds (see Fig. 9.12) are approximately $N^+ = 18.5$, $N^{++} = 25$. The dispersion of the congestion distribution for $N^+ - 1 < N = 20 < N^{++} - 1$ appears in Fig. 9.13 as more important *in relation* to N than the dispersions for $N = 17 < N^+ - 1$ and for $N = 25 > N^{++} - 1$. In these latter cases no unstable solution B exists and one can verify that the corresponding distributions of Fig. 9.13 exhibit a single maximum equal to $n_A = 0$ and $n_C = 11$, respectively.

The evolution of the *variance* of the congestion in function of N shows this phenomenon more clearly. The variance $\sigma^2(N)$ of the distribution $\{\pi_2(n_1 \mid N)\}$, $0 \leqslant n_1 \leqslant N$,

$$\sigma^2(N) = \sum_{n_1=0}^{N} [E(N) - n_1]^2 \pi_2(n_1 \mid N),$$

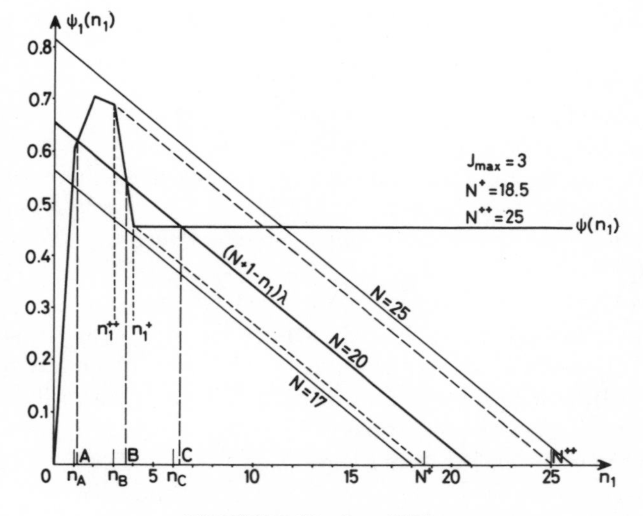

FIGURE 9.12 Instabilities.

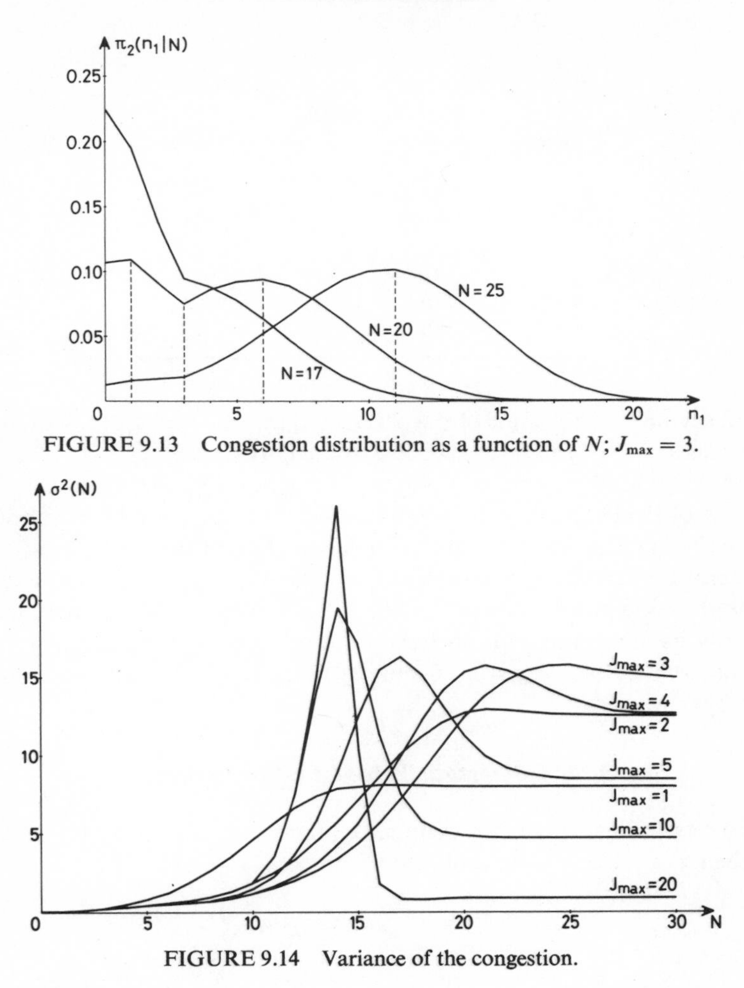

FIGURE 9.13 Congestion distribution as a function of N; $J_{max} = 3$.

FIGURE 9.14 Variance of the congestion.

where $E(N)$ is used to denote the mean congestion, is shown in Fig. 9.14 as a function of N and for various values of J_{max}.

This figure shows that for each value of J_{max} there is a value of N, we shall call it S_V, that maximizes $\sigma^2(N)$. It is remarkable that this value S_V is, for any J_{max}, situated in the interval $[(N^+ - 1), (N^{++} - 1)]$; this can be verified with Table 9.1.

This sudden increase of the variance of the congestion in the region $[(N^+ - 1), (N^{++} - 1)]$ is, of course, accompanied by a corresponding increase of the variance of the response time. Such increases had been experimentally observed but left unexplained in a simulation study of the system ESOPE (Betourne and Krakowiak, 1972).

TABLE 9.1

J_{max}	$N^{+}-1$	S_V	$N^{++}-1$
1	9	17	20
2	14	22	24
3	17	25	25
4	16	21	25
5	13	17	25
10	10	14	25
20	10	14	25

The asymptotic behavior of the variance shown by Fig. 9.14 for values of N larger than $N^{++}-1$ will be discussed in the next section.

The relationship we have discussed here between the variance and the extrema of the congestion distribution, and the solutions of the type A, B, C of Eq. (9.23) backs up our intuitive understanding of the stochastically stable and unstable operating points of the system: the nearer to a stable value a congestion, the higher is its probability of occupancy. Alternatively, unstable congestions have minimum probability of occupancy: they are only "passed through" when the congestion fluctuates randomly from one stable region to the other.

9.12 Asymptotic Behavior of the Congestion

The mean congestion in the system, i.e., the mean value of n_1, the number of jobs being or waiting to be multiprogrammed, is given by

$$E(N) = \sum_{n_1=0}^{N} n_1 \times \pi_2(n_1 \mid N). \tag{9.24}$$

In Fig. 9.15, this mean system congestion $E(N)$ is plotted against the number N of active user terminals, $1 \leqslant N \leqslant 20$, for different values of J_{max}.

9.12.1 The first thing we note is that, as $N \to \infty$, $E(N)$ approaches an *asymptote*, which is a function of the value of J_{max}. This can be explained as follows. In the steady state we may equate the mean input rate and the mean service rate; the mean input rate is equal to

$$\sum_{n_1=0}^{N} \pi_2(n_1 \mid N) \times (N-n_1)\lambda = [N-E(N)]\lambda. \tag{9.25}$$

Let us use $\bar{\psi}(N)$ to denote the mean service rate

$$\bar{\psi}(N) = \sum_{n_1=0}^{N} \pi_2(n_1 \mid N)\psi(n_1). \tag{9.26}$$

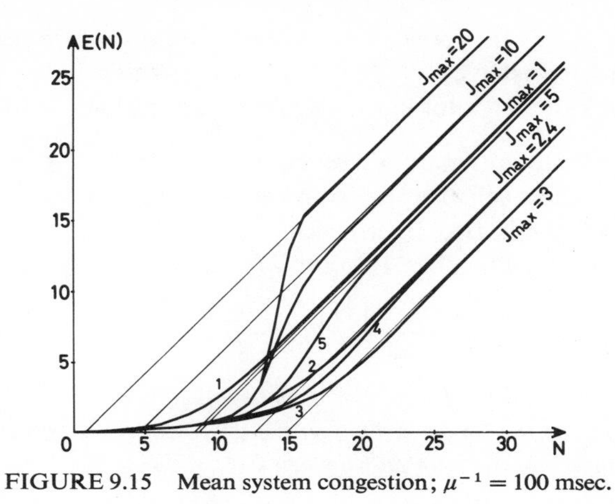

FIGURE 9.15 Mean system congestion; $\mu^{-1} = 100$ msec.

Equating (9.25) and (9.26), we find the balance equation

$$E(N) = N - \sum_{n_1=0}^{N} \pi_2(n_1 \mid N) \frac{\psi(n_1)}{\lambda} = N - \frac{\bar{\psi}(N)}{\lambda}. \tag{9.27}$$

Moreover, from Eq. (9.20) for $0 \leqslant n_1 \leqslant N$, we derive

$$\pi_2(n_1 \mid N) \leqslant \psi(n_1+1)/(N-n_1)\lambda. \tag{9.28}$$

Expression (9.28) indicates that for a fixed n_1, $\pi_2(n_1 \mid N)$ approaches zero as $N \to \infty$. In particular, as $N \to \infty$, all probabilities $\pi_2(n_1 \mid N)$ approach zero for $n_1 < J_{\max}+1$. Since $\psi(n_1) = \psi(J_{\max}+1)$ for $n_1 \geqslant J_{\max}+1$, from (9.26) we obtain

$$\lim_{N \to \infty} \bar{\psi}(N) = \psi(J_{\max}+1), \tag{9.29}$$

and from (9.27)

$$\lim_{N \to \infty} [E(N)-N] = -\psi(J_{\max}+1)/\lambda. \tag{9.30}$$

This proves that *the mean service rate* $\bar{\psi}(N)$ *has an asymptote* $\bar{\psi}(N) = \psi(J_{\max}+1)$ *and that the mean congestion* $E(N)$ *has an asymptote of slope* 1 *that intersects the N-axis at abcissa* $\psi(J_{\max}+1)/\lambda$.

Let us note also that $\psi(J_{\max}+1)/\lambda = N - C$ (see Fig. 9.12, for example). Thus,

$$\lim_{N \to \infty} E(N) = C \approx n_C.$$

This is in agreement with our previous results. When $N > N^{++}$, then n_C remains the only stable congestion; it is also the congestion of maximum probability. As $N \to \infty$, also $n_C \to \infty$, and the probabilities for the small congestions at which the service rate is not constant become negligible. Thus the system tends to behave as a system of constant service rate $\psi(J_{max}+1)$. We shall come back to this point in the section on saturation.

9.12.2 The second observation we can make on Fig. 9.15 is that these asymptotes are approached either from below or from above, and in one case ($J_{max} = 20$) even with two intersection points. To a certain extent, such behaviors are also predictable. From (9.27), (9.29), and (9.30) we obtain

$$E(N) - \left[N - \frac{\psi(J_{max}+1)}{\lambda}\right] = -\frac{1}{\lambda}[\bar{\psi}(N) - \psi(J_{max}+1)]. \tag{9.31}$$

Thus, *the intersections of $E(N)$ with its asymptote have the same N-value as the intersections of the mean service rate $\bar{\psi}(N)$ with its asymptote, and $E(N)$ approaches its asymptote from above when $\bar{\psi}(N)$ approaches its asymptote from below and vice versa.*

It is harder to predict the asymptotic behavior of $E(N)$ from the behavior of the service rate $\psi(n_1)$. However, we can prove that *if $\psi(n_1)$ is a nondecreasing function of n_1, i.e., if no time slice is enforced and $J_{max} \leqslant J_0$, then $E(N)$ has no finite intersection with its asymptote, this one being approached from above.*

PROOF If $\psi(n_1)$ is nondecreasing, we have $\psi(J_{max}+1) = \max_{n_1 \geqslant 0}[\psi(n_1)]$. Then it is clear that for any finite N,

$$\bar{\psi}(N) < \psi(J_{max}+1),$$

since $\psi(0) = 0$ and $\pi_2(0 \mid N)$ is never identically zero for N finite. Therefore, $\bar{\psi}(N)$ has no finite intersection with its asymptote and approaches this asymptote from below. Thus, $E(N)$ has no finite intersection with its asymptote either, and approaches it from above.

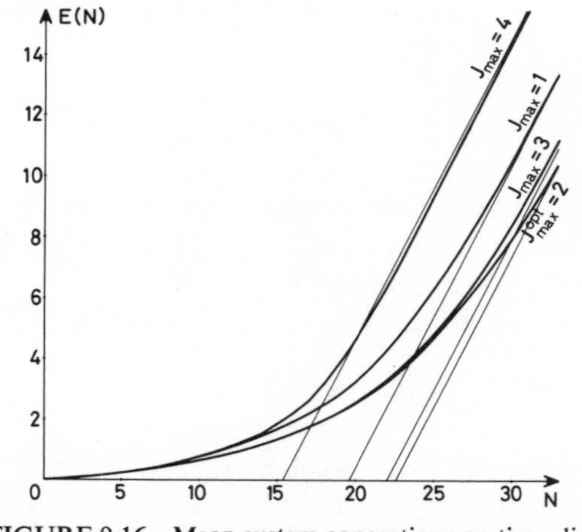

FIGURE 9.16 Mean system congestion; no time slice.

This latter case is illustrated by Fig. 9.16, which shows $E(N)$ when no time slice is enforced ($\mu^{-1} = \phi^{-1}$). In this case, $\psi(n_1)$ is nondecreasing for $J_{max} \leqslant J_0 = 2$. We can observe on Fig. 9.16 that there is no finite intersection of $E(N)$ with its asymptote in this case, and one intersection for only $J_{max} = 4$.

9.12.3 A more general property of the behavior of $E(N)$ as a function of $\psi(n_1)$ can be conjectured. From expressions (9.18) we derive

$$\frac{\pi_2(n_1 \mid N)}{\pi_2(n_1 \mid N-1)} = \frac{N}{N-n_1} \frac{\pi_2(0 \mid N)}{\pi_2(0 \mid N-1)}.$$

We can rewrite $\pi_2(0 \mid N)$ so that

$$\frac{\pi_2(0 \mid N)}{\pi_2(0 \mid N-1)} = \frac{\sum_{j=0}^{N-1} \lambda^j (N-1)!/(N-j-1)! \prod_{k=1}^{j} \psi(k)}{\sum_{j=0}^{N} \lambda^j N!/(N-j)! \prod_{k=1}^{j} \psi(k)}$$

$$= \frac{1}{N} \frac{\sum_{j=0}^{N-1} \lambda^j/(N-j-1)! \prod_{k=1}^{j} \psi(k)}{1/N! + \sum_{j=0}^{N-1} \lambda^{j+1}/(N-j-1)! \prod_{k=1}^{j+1} \psi(k)}.$$

Defining $\psi max(N) = \max_{1 \leqslant j \leqslant N} [\psi(j)]$, we obtain

$$\pi_2(0 \mid N)/\pi_2(0 \mid N-1) < \psi max(N)/N\lambda,$$

and therefore

$$\pi_2(n_1 \mid N) < [\psi max(N)/(N-n_1)\lambda]\, \pi_2(n_1 \mid N-1).$$

We conclude that for any n_1 and N such that the input rate $(N-n_1)\lambda$ is larger than the maximum service rate $\psi max(N)$, probability $\pi_2(n_1 \mid N)$ decreases as N increases. This has the consequence that $\bar{\psi}(N)$ *is an increasing function of N provided that* $\psi(n_1)$ *is a nondecreasing function of* n_1 *for* $0 \leqslant n_1 \leqslant N$. From this we can conjecture that $E(N)$ *has no more intersections with its asymptote than* $\psi(n_1)$ *has with the line* $\psi(n_1) = \psi(J_{max}+1)$.

SKETCH OF A PROOF For $N = 0$, $\bar{\psi}(0) = \psi(0) = 0 < \psi(J_{max}+1)$. Thus, as N increases, we shall have $\bar{\psi}(N) \geqslant \psi(J_{max}+1)$ only if congestions n_1 exist such that $\psi(n_1) \geqslant \psi(J_{max}+1)$. When $\bar{\psi}(N) \geqslant \psi(J_{max}+1)$, we know that $\bar{\psi}(N)$ continues to increase as long as $\psi(n_1)$, $0 \leqslant n_1 \leqslant N$, is nondecreasing. Thus, as N increases, $\bar{\psi}(N)$ will decrease only if $\psi(n_1)$ does, and since for all n_1 such that $(N-n_1)\lambda > \psi max(N)$, probabilities $\pi_2(n_1 \mid N)$ tend uniformly to zero as N increases, $\bar{\psi}(N)$ will decrease below $\psi(J_{max}+1)$ only if congestions $n_1' > n_1$ exist such that $\psi(n_1') < \psi(J_{max}+1)$. Therefore, the difference $\bar{\psi}(N) - \psi(J_{max}+1)$, and thus the difference (9.31) cannot change sign unless the difference $\psi(n_1) - \psi(J_{max}+1)$ does.

One can, for example, verify that $E(N)$ has two finite intersections with asymptote $J_{\max} = 20$ in Fig. 9.15 and that in Fig. 9.9 the corresponding function $\psi(n_1)$ has three intersections with the line $\psi(n_1) = \psi(21)$. The many intersections of $\psi(n_1)$ with the line $\psi(J_{\max}+1)$ result here from the slight $\sigma^*(J_{\max})$ increases discussed in Remark 9.9.2.

Of course, $E(N)$ may have fewer intersections with its asymptote than $\psi(n_1)$ with the horizontal $\psi(n_1) = \psi(J_{\max}+1)$. This is typically illustrated by the asymptote $J_{\max} = 3$ in Fig. 9.16 and also by the asymptote $J_{\max} = 1$ on Fig. 9.15. There is no finite intersection in these two cases, although $\psi(n_1)$ takes twice the value $\psi(J_{\max}+1)$, (see Fig. 9.9). If we approximate $\overline{\psi}(N)$ by $\psi(n_A)$, n_A being the stable congestion of maximum probability, this can be partly explained by the fact that $\psi(n_A)$, for any N, never exceeds $\psi(J_{\max}+1)$ in these two cases.

9.12.4 For large values of N, the limiting behavior of $\sigma^2(N)$, the variance of the congestion distribution, can be simply predicted as follows.

For $N \to \infty$, we already know that the system tends to behave as a system with constant service rate $\psi(J_{\max}+1)$, and that the mean congestion $E(N)$ increases with N beyond limit. Therefore, in the dual system, the distribution of the number $N-n_1$ of thinking users must approach the distribution of the congestion in a simple system with infinite input population and constant Poisson input rate $\psi(J_{\max}+1)$, and with infinitely many available servers (terminals) and an output rate $(N-n_1)\lambda$. This distribution is Poisson with parameter $\psi(J_{\max}+1)/\lambda$ (Feller, 1968, XVII.7). Thus, we have

$$\lim_{\substack{N\to\infty,\, n_1\to\infty \\ (N-n_1)\,\mathrm{const}}} \Pr\{(N-n_1) \text{ thinking users}\}$$

$$= \lim_{\substack{N\to\infty,\, n_1\to\infty \\ (N-n_1)\,\mathrm{const}}} \pi_2(n_1 \mid N)$$

$$= \frac{1}{(N-n_1)!}\left[\frac{\psi(J_{\max}+1)}{\lambda}\right]^{N-n_1} \exp\left[-\frac{\psi(J_{\max}+1)}{\lambda}\right], \qquad 0 \leqslant n_1 \leqslant N.$$

The mean and variance of this distribution are equal to $\psi(J_{\max}+1)/\lambda$, so that we have

$$\lim_{N\to\infty} \mathbb{E}\{N-n_1\} = \lim_{N\to\infty} \sigma^2\{N-n_1\} = \psi(J_{\max}+1)/\lambda,$$

or, as we already know,

$$\lim_{N\to\infty} [E(N)-N] = -\psi(J_{\max}+1)/\lambda$$

and

$$\lim_{N\to\infty} \sigma^2(N) = \psi(J_{\max}+1)/\lambda. \tag{9.32}$$

This limiting behavior of the variance can be observed on Fig. 9.14.

9.12.5 ***Minimizing the Mean Congestion*** We have now a sufficient understanding of the different mechanisms influencing the behavior of the mean congestion to determine an optimal choice of the maximum degree of multiprogramming J_{max} that minimizes the congestion. We know from Eq. (9.27) that

$$E(N) = N - \bar{\psi}(N)/\lambda,$$

where $\bar{\psi}(N)$ is the mean service rate. Thus, for a fixed λ and a given N, *the minimum mean congestion will be obtained for a maximum mean service rate.* Our two degrees of freedom to maximize $\bar{\psi}(N)$ are J_{max} and the length of the execution interval. Two cases must be distinguished.

A. For $N > N^{++}$, the system behaves almost as a system of constant service rate $\psi(J_{max}+1)$, so that $\bar{\psi}(N) \approx \psi(J_{max}+1)$. But

$$\psi(J_{max}+1) = \phi \times \sigma^*(J_{max}),$$

and we know from Section 9.9 that $\sigma^*(J_{max})$ is maximum when the execution interval is chosen as long as possible and when $J_{max} = J_{max}^{opt}$. Therefore, in this first case, these are the conditions that minimize the mean congestion. If the execution interval can be chosen infinite (no time slicing required), then $J_{max}^{opt} = J_0$, $\psi(J_{max}+1) = \psi(J_{max}) = \psi(J_0)$, and the mean congestion is at an absolute minimum. There is another way to show that for large N values minimizing the mean congestion amounts to maximizing $\psi(J_{max}+1)$: the asymptote of minimal congestions on a diagram $[E(N); N]$, such as Fig. 9.15, is clearly the asymptote of maximum abscissa $\psi(J_{max}+1)/\lambda$.

B. When $N < N^{++}$, a different situation may arise. Again the minimum mean congestion will be obtained for the values of J_{max} and of the execution interval that maximize $\bar{\psi}(N)$. However, for a given execution interval, the optimum does not correspond necessarily to J_{max}^{opt}. Suppose we have the situation illustrated by Fig. 9.17. It is clear from this figure that $\bar{\psi}(N_1)$ is

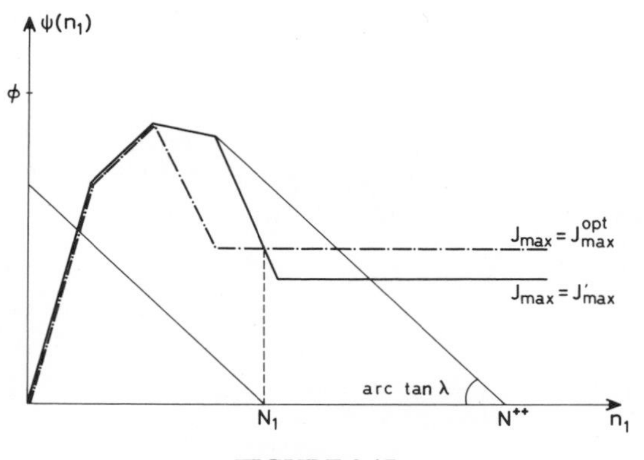

FIGURE 9.17.

larger for $J_{max} = J'_{max}$ than for $J_{max} = J_{max}^{opt}$. The mean congestion will therefore be minimum for $J_{max} = J'_{max} < J_{max}^{opt}$ in the vicinity of N_1. A similar situation is encountered in our example: computation results have shown that in the region $N = 13, 12, 11, \ldots, E(N)$ is smaller for $J_{max} = 5$ than for $J_{max} = 2$, although $\psi(J_{max}+1)$ is larger for $J_{max} = 2$ than for $J_{max} = 5$ (see Fig. 9.9).

In such cases and for a given execution interval, the optimal maximum degree of multiprogramming can only be determined as the value of J_{max} that maximizes $\bar{\psi}(N)$ for the relevant range of N values; a graphical determination will often be possible from a balance diagram such as Fig. 9.17.

9.12.6 ***Optimal Time Slice*** It follows from the above discussion that minimizing the mean congestion implies in particular to choose the longest possible execution interval. Suppose now that it would be desirable that small jobs requiring less than a given amount of processor time be guaranteed service within a maximum waiting time. The way to achieve this is to make use of a time slice, calculated in function of this prerequisite maximum waiting time. Shortening this time slice will, up to a certain point, improve this waiting time, but at the same time increase the mean system congestion. An optimal time slice must therefore exist. The determination of this optimum requires a similar but more elaborate model than ours, in which jobs would be classified according to their demand in processor time.

9.13 Saturation

Figure 9.15 shows that the slope of the mean congestion $E(N)$ is less than unity for small values of N and, as N increases, may or may not sharply exceed unity, depending on the value of J_{max}; for larger values of N, the slope becomes asymptotically equal to unity. Such variations can be explained by the considerations made in the previous sections and can be related to an intuitive concept of saturation.

9.13.1 Let us use $\Delta E(N)$ and $\Delta\bar{\psi}(N)$ to denote the first differences of $E(N)$ and $\bar{\psi}(N)$ consecutive to an increment ΔN of N:

$$\Delta E(N) = E(N+\Delta N) - E(N), \qquad \Delta\bar{\psi}(N) = \bar{\psi}(N+\Delta N) - \bar{\psi}(N).$$

From Eq. (9.27), we obtain

$$\frac{\Delta E(N)}{\Delta N} = 1 - \lambda^{-1}\frac{\Delta\bar{\psi}(N)}{\Delta N}. \tag{9.33}$$

Hence, we have

$$\begin{aligned} \Delta E(N)/\Delta N &< 1 \quad \text{iff} \quad \Delta\bar{\psi}(N)/\Delta N > 0, \\ &\geqslant 1 \quad \text{iff} \quad \Delta\bar{\psi}(N)/\Delta N \leqslant 0. \end{aligned} \tag{9.34}$$

Clearly, the slope of $E(N)$ remains less than unity as long as the slope of $\bar{\psi}(N)$ is positive, reaches unity when $\bar{\psi}(N)$ is maximum, and exceeds unity once $\bar{\psi}(N)$ has a negative slope.

When the slope of the mean congestion in the system is equal to or above unity, each additional active user terminal adds, on the average, at least one more unit to the congestion, and at least one more mean processing time to the mean response time. In other words, each additional user interferes completely with the others. This state of affairs corresponds to an intuitive notion of *saturation*: we can define the *saturation load* of the system, say S, as the largest value of N such that for all N less than or equal to this value, the slope of $E(N)$ is less than unity. As shown by Eq. (9.33), S is exactly *the first value of N that maximizes the mean service rate* $\bar{\psi}(N)$.

9.13.2 An exact and general determination of S requires an analytical definition of the function $\bar{\psi}(N)$, but in all generality, this definition does not exist: $\psi(n_1)$ can be an arbitrary function known only numerically, and in this case $\bar{\psi}(N)$ can only be obtained by computation.

However, S is easy to find if $\psi(n_1)$ is nondecreasing for all n_1. In this case we know from Section 9.12.3 that $\bar{\psi}(N)$ is increasing for all N, and thus $S = \infty$.

In all the other cases, when $\psi(n_1)$ is decreasing for some n_1, only a general approximation of S, say S_A, can be obtained. We make the following reasoning. As long as $N < N^+ - 1$, there is only one stable congestion n_A that is also a congestion of maximum probability. We can therefore expect the variations of $\bar{\psi}(N)$ to follow those of $\psi(n_A)$. More precisely, we can assume that for $N < N^+ - 1$, $\bar{\psi}(N)$ reaches a maximum if $\psi(n_A)$ does, with $n_A = N + 1 - \psi(n_A)/\lambda$. Let us then use n_0 to denote the congestion for which $\psi(n_1)$ is maximum; we have

$$n_0 = \min(J_0, J_{\max}).$$

Thus, defining

$$N_0 = n_0 + \psi(n_0)/\lambda,$$

we may approximate S by *integer*$(N^+ - 1)$ if $N^+ < N_0$ (Fig. 9.18) and by *integer*$(N_0 - 1)$ otherwise (Figs. 9.19 and 9.20). Therefore,

$$S_A = \textit{integer}[\min(N_0, N^+) - 1]. \tag{9.35}$$

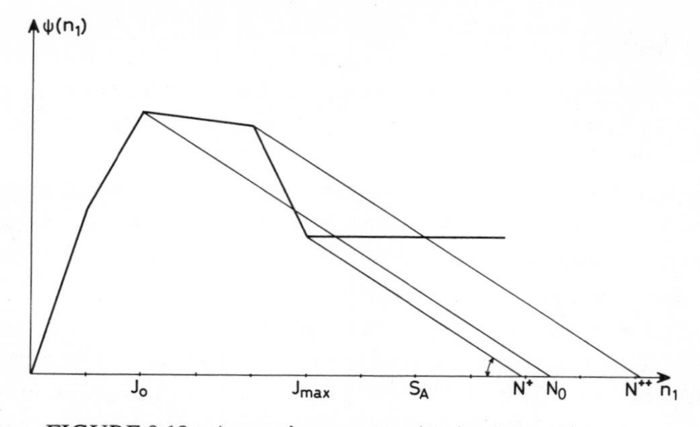

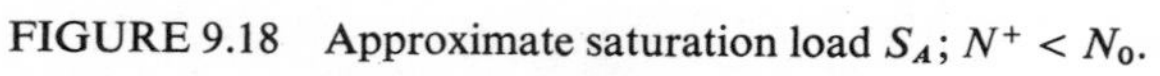

FIGURE 9.18 Approximate saturation load S_A; $N^+ < N_0$.

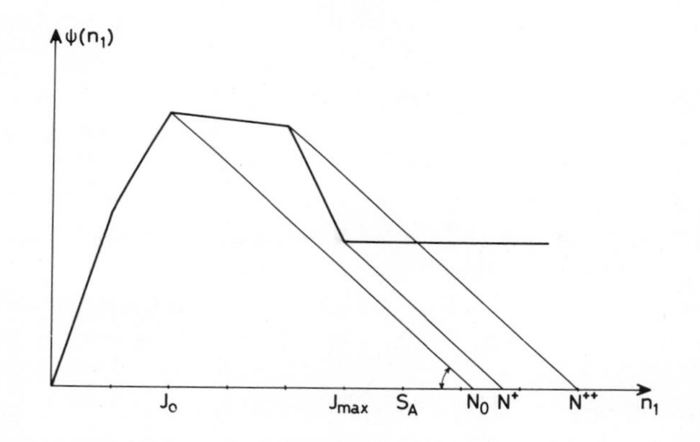

FIGURE 9.19 Approximate saturation load S_A; $N_0 < N^+$.

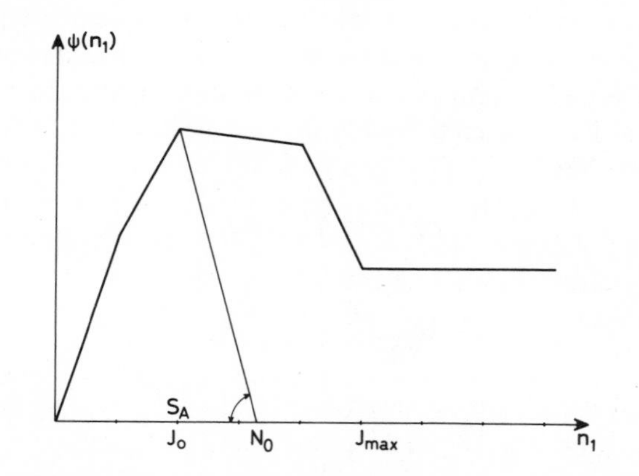

FIGURE 9.20 Approximate saturation load S_A; no instability.

Let us immediately remark that once $N \geqslant N^+ - 1$, the distribution $\{\pi_2(n_1 \mid N\}_{n_1=0}^{N}$ has two maxima, and nothing very accurate can be said on an approximation of $\bar{\psi}(N)$ in terms of $\psi(n_A)$ or $\psi(n_C)$. Nevertheless, as $\psi(n_C) < \psi(n_A)$, one must expect that in general $\bar{\psi}(N)$ will eventually decrease for N beyond $N^+ - 1$.

Summing up, *when $\psi(n_1)$ is decreasing for some n_1, we approximate the saturation load S by the largest integer N such that (i) there is only one stable solution A of* (9.23) *and (ii) this solution lies in a region where $\psi(n_1)$ is increasing. Otherwise, when $\psi(n_1)$ is nondecreasing, S is infinite.*

9.13.3 In general, S_A will be smaller than the exact saturation load S; this can be verified in Table 9.2 in which we compare the values of S and S_A corresponding to the curves $E(N)$ displayed on Fig. 9.15. Two reasons can be given, depending on the value taken by S_A:

TABLE 9.2

J_{max}	S	S_A	S_V	S_K
1	15	9	17	17
2	19	14	22	25
3	22	17	25	30
4	18	16	21	26
5	14	13	17	19
10	12	10	14	14
20	12	10	14	13

(1) $S_A = integer(N_0 - 1)$. Then $\bar{\psi}(N)$ might increase beyond S_A, even though $\psi(n_A)$, an approximation of $\bar{\psi}(N)$, decreases. Indeed, although they lie in the decreasing part of $\psi(n_1)$, the values of $\psi(n_1)$ around $\psi(n_A)$ supplant in $\bar{\psi}(N)$, as N increases, the much smaller values of $\psi(n_1)$ around $\psi(0) = 0$; these latter values indeed have a probability approaching zero like $\pi_2(0 \mid N)$, almost as fast as $1/N!$ [see Eq. (9.18)].

(2) $S_A = integer(N^+ - 1)$. Then S_A is the very first value of N for which a second stable intersection n_C exists. In general, however, $\pi_2(n_C \mid N)$ remains very small when N is only slightly above S_A; this implies that the small value $\psi(n_C)$ hardly influences $\bar{\psi}(N)$, which thus continues for a while to vary approximately like $\psi(n_A)$, and to increase beyond S_A.

In practice, however, S_A is a safe and acceptable approximation of S. Table 9.2 shows that the largest deviations between S_A and S occur, fortunately, for the small values of J_{max} at which the phenomenon of saturation is the least pronounced (this will be clarified in Section 9.13.5). Moreover, the approximation of $\bar{\psi}(N)$ by $\psi(n_A)$, and thereby the approximation of S by S_A, improves when the variance of the distribution $\{\pi_2(n_1 \mid N)\}_{n_1=0}^{N}$ is small. This, as indicated by (9.20), occurs when λ is large.

9.13.4 Let us also remark that instabilities and saturation are phenomena that are tightly related to each other. Once $N \geqslant N^+ - 1$, the distribution $\{\pi_2(n_1 \mid N)\}_{n_1=0}^{N}$ has two distinct maxima, n_A and n_C, responsible for the increase of the variance $\sigma^2(N)$. It is, however, only when N is equal to S_V that the probability $\pi_2(n_C \mid N)$ is sufficiently large to maximize $\sigma^2(N)$. But it is also this sufficiently large probability $\pi_2(n_C \mid N)$ that, as we have seen above, causes $\bar{\psi}(N)$ to decrease and the saturation point to be reached. One can verify in Table 9.2 that S_V and S remain close to each other.

9.13.5 ***Avalanche Effect*** Once $N \geqslant N^+ - 1$, intersections B and C exists as well as A (see Fig. 9.12), and $E(N)$ must rise abruptly up to an average over the range of values between the two stable values n_A and n_C. The smaller $\psi(n_C)$, the larger $|-\Delta\bar{\psi}(N)/\Delta N|$ and $\Delta E(N)/\Delta N$ will be. Such sharp nonlinearities in the vicinity of $N^+ - 1$ can be observed in Fig. 9.15 for $J_{max} = 5, 10, 20$; they can be understood as the consequence of some kind of positive feedback that takes place as soon as the congestion exceeds the unstable congestion n_B. In this region an increase of the congestion results in an increase of the positive difference between the input and the output rates, which in turn accentuates the initial congestion increase. The maximum imposed on the degree of multiprogramming acts, of course, as a barrier to this *avalanche effect*: the system service rate never decreases below $\psi(J_{max}+1)$, no matter how large the congestion becomes.

This avalanche effect is best illustrated by Table 9.3 in which some representative values of the mean congestion $E(20)$ are plotted against J_{max}. These values clearly indicate how the system may become overcrowded when J_{max} is not chosen close enough to J_{max}^{opt} so as to reduce to a minimum this avalanche effect and the accompanying instabilities.

TABLE 9.3

J_{max}	$E(20)$
1	11.9
2	7.3
$J_{max}^{opt} =$ 3	4.9
4	6.6
10	15.2
20	19.1

9.13.6 ***Extension of Kleinrock's Definition*** Kleinrock (1968) had studied the phenomenon of saturation in queueing systems with *constant* service rate and finite input population. If the constant service rate is ψ, then the mean service rate of such systems becomes

$$\bar{\psi}(N) = \psi[1-\pi_2(0\,|\,N)],$$

and since $\pi_2(0\,|\,N)$ tends uniformly to zero when N increases, $\bar{\psi}(N)$ is an increasing function of N. Therefore the slope of the mean congestion $E(N)$ is *always* smaller than unity in such systems, and Kleinrock had necessarily to take another approach to define saturation in these systems.

Replacing each service time by exactly its average ψ^{-1} and each user reaction time by exactly its average λ^{-1}, he obtained a deterministic system that can handle a maximum of

$$N^* = \frac{\psi^{-1}+\lambda^{-1}}{\psi^{-1}} = 1+\frac{\psi}{\lambda} \tag{9.36}$$

consoles without mutual interference and queueing delay among the users. This provided Kleinrock with an intuitive basis on which to define N^* as the saturation load of an N-console stochastic queueing system with constant service rate. For $N < N^*$ the mean congestion tends to increase very slowly since each user tends to request service while the others are thinking. For $N > N^*$, the mean congestion approaches and follows by excess an asymptote of slope 1, each additional user interfering, on the average, completely with the others. This asymptote intersects the line $E(N) = 1$ at precisely abscissa N^*.

Our analysis is in agreement with this last property; we can verify with Eq. (9.30) that our $E(N)$ asymptotes intersect the line $E(N) = 1$ at abscissas that are equal to $\psi(J_{\max}+1)/\lambda$. This is in accordance with the fact that the asymptotic behavior of our system is similar to that of a system with constant service rate $\psi(J_{\max}+1)$.

However, Kleinrock's definition of saturation is not adequate for our system. Based on the assumption that the service rate is constant, it cannot take into account the variations of $\psi(n_1)$ for $n_1 \leqslant J_{\max}+1$.

It is yet possible (Courtois and Vantilborgh, 1975) to generalize Kleinrock's notion of saturation to systems with congestion-dependent service rate. With probability $\pi_2(n_1\,|\,N)$, there are η_1 users in the system and the service rate is $\psi(n_1)$, so that following Kleinrock's argument, $[\psi(n_1)/\lambda]+1$ users can, on the average, be served without mutual interference. Thus, with respect to this ideal average situation, there is a users excess equal to

$$n_1 - \left[\frac{\psi(n_1)}{\lambda}+1\right];$$

an average of this users excess is then given by

$$e(N) = \sum_{n_1=0}^{N} \left[n_1 - \frac{\psi(n_1)}{\lambda} - 1 \right] \pi_2(n_1 \mid N),$$

or

$$e(N) = E(N) - \frac{1}{\lambda}\bar{\psi}(N) - 1;$$

and since $E(N) = N - \bar{\psi}(N)/\lambda$, we obtain

$$e(N) = N - 2\frac{\bar{\psi}(N)}{\lambda} - 1 = 2E(N) - N - 1. \tag{9.37}$$

Therefore, a natural extension of Kleinrock's intuitive notion of saturation is to consider the system as being saturated once $e(N) > 0$. In this case, the saturation load of the system, say S_K, is the largest value such that for all N smaller than or equal to this value, we have

$$e(N) \leqslant 0, \qquad \text{i.e.} \qquad N \leqslant 1 + 2\bar{\psi}(N)/\lambda, \qquad \text{i.e.} \qquad E(N) \leqslant \tfrac{1}{2}(N+1).$$

Figures 9.21 and 9.22 show how S_K can be graphically determined.

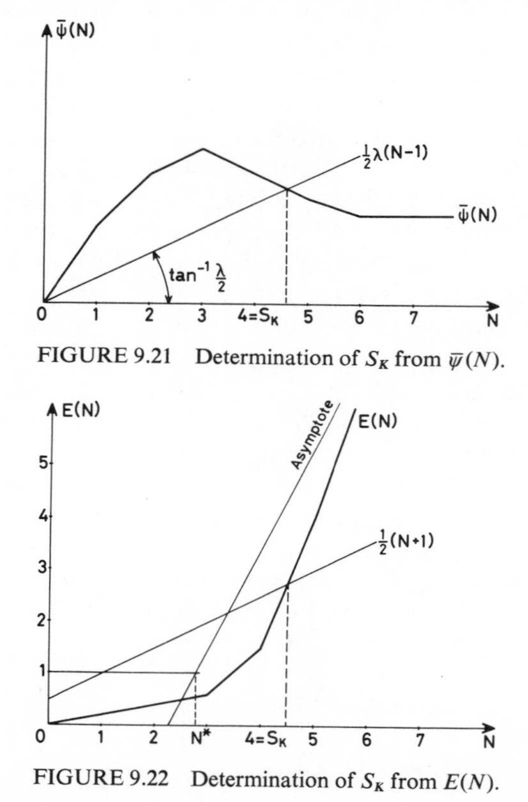

FIGURE 9.21 Determination of S_K from $\bar{\psi}(N)$.

FIGURE 9.22 Determination of S_K from $E(N)$.

Of course, the determination of S_K also requires the full knowledge of $\bar{\psi}(N)$. Table 9.2 gives the values of S_K that correspond to the curves $E(N)$ displayed on Fig. 9.15. This table shows that S_K is larger than S for these curves. This can be clearly understood from the way S_K is graphically determined: $\Delta\bar{\psi}(N)/\Delta N$ can be negative or null for N smaller than S_K; the larger λ, the closer S_K will be to S.

9.13.7 Finally, Table 9.2 shows that the larger $J_{\max}$, the closer to each other the critical loads S, S_A, S_V, S_K are. As we know, the larger $J_{\max}$, the more severe the phenomena we have studied in this section are also. The proximity of these critical loads must be interpreted as an indication that these loads identify only the various consequences of the same phenomenon, which causes a sudden deterioration of the general performances of the system.

9.14 System Response Time

The mean response time $W(N)$ of the system, i.e., the mean time spent by a program being or waiting to be multiprogrammed, can be obtained from Little's formula (5.16):

$$W(N) = E(N)/\bar{\psi}(N), \tag{9.38}$$

which, by replacing $E(N)$ by expression (9.27), yields

$$W(N) = \frac{N}{\bar{\psi}(N)} - \frac{1}{\lambda}. \tag{9.39}$$

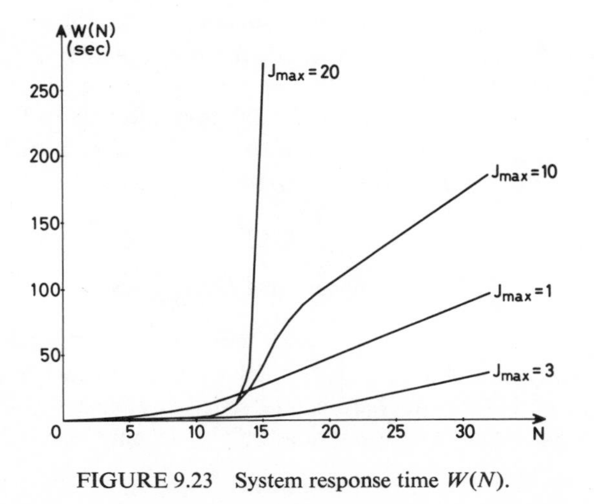

FIGURE 9.23 System response time $W(N)$.

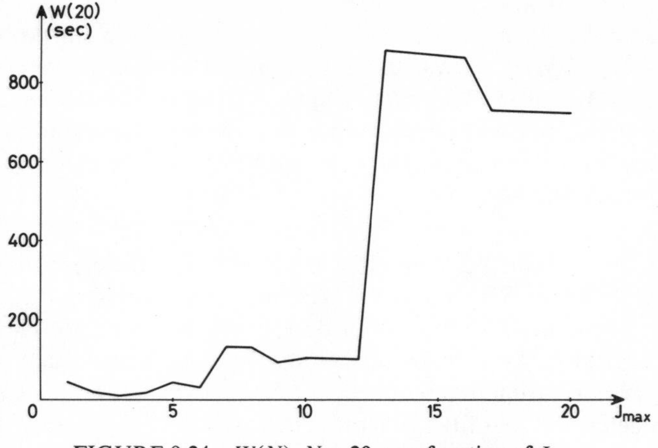

FIGURE 9.24 $W(N)$, $N = 20$, as a function of J_{max}.

In Fig. 9.23, $W(N)$ is plotted as a function of N for $J_{max} = 1, 3, 10$, and 20. As in the case of the mean congestion, sharp nonlinearities appear once N exceeds the corresponding saturation load, which for these J_{max} values is given by $N^+ - 1$. For larger values of N, the system tends to behave as a system with congestion-independent service rate $\psi(J_{max}+1)$, so that we have

$$\lim_{N\to\infty} W(N) = \frac{N}{\psi(J_{max}+1)} - \frac{1}{\lambda}. \tag{9.40}$$

In particular, for the limiting slope of $W(N)$ as a function of N we have

$$\lim_{N\to\infty} \Delta W(N)/\Delta N = \psi(J_{max}+1)^{-1} = [\phi \times \sigma^*(J_{max})]^{-1}.$$

Therefore, the smaller $\sigma(J_{max})$ and $\psi(J_{max}+1)$, the steeper this limiting slope is.

The fluctuations of $W(N)$ as a function of J_{max} for N fixed are shown in Fig. 9.24; the many discontinuities of $W(N)$ as a function of J_{max} correspond to the discontinuities of $\sigma^*(J_{max})$ (cf. Fig. 9.9).

Finally we can deduce from Eq. (9.39) that the value of the maximum degree of multiprogramming that tends to minimize the congestion is also the value that will minimize the mean response time, since this value maximizes the mean service rate $\bar{\psi}(N)$.

9.15. Conclusions and Open Questions

The asymptotic properties of near-complete decomposability made it possible to analyze, within a single framework, the influence upon the overall system performance of many parameters, specifying the demand upon various resources of the system and specifying the policies that govern the allocation of these resources. Without the distinction between short- and long-term equilibrium, this analysis would have been much more difficult. In particular, the conditions of existence of phenomena such as saturation and instabilities would have been much harder to isolate. In this respect, we believe that a load diagram such as Fig. 9.12 is an easily obtained and a significant aid for the designer of a time-sharing multiprogramming system; it allows him to predict easily for wide ranges of values of many different parameters the type of behavior that can be expected from his system; it also allows him to derive optimal values for these parameters easily.

The analysis has, however, not been pursued to the end; the influence of many parameters has not yet been studied. For example, all the material is there to analyze the influence of the characteristics of the auxiliary memory upon the global system response. Moreover, the role played by the time slice has been only partially studied; we have studied its influence on the page fault rate but have not considered how it affects the time spent in the system by a particular class of jobs having a given execution time. Such an analysis might lead to the definition of an optimal time slice that minimizes the response time for that class of jobs.

Besides, new parameters could be added to the model without too much difficulty. We have in mind, for instance, the modeling of different page replacement policies, taking into account their implementation costs in terms of space and time. Such a comparative study could shed some light on the still much debated question of the real gain achieved by these policies at the level of the global performances of the system. This could be done by using for the page fault rate the values yielded by the models of Chapter VIII.

CHAPTER X

Hierarchical System Design

By way of conclusion, in this closing chapter we make a few conjectures that in our opinion are important, but that are still, unfortunately, at the stage of half-baked ideas. These conjectures are drawn from striking analogies that exist between, on the one hand, the concept of aggregate in nearly completely decomposable systems and, on the other hand, the less precisely defined notion of building block, or module, so frequently invoked in computer system hardware and software architecture.

The technique of aggregation allows us to (i) break a system up into a small number of subsystems, (ii) evaluate the interactions within the subsystems as though interactions among subsystems did not exist, and (iii) evaluate the interactions among subsystems without regard to the interactions within subsystems. We may expect this evaluation technique to bear some resemblance to the design technique whereby a complex system may be assembled from subsystems that have been designed independently. In particular, we may expect the criteria that indicate which variables should be aggregated also to be useful in specifying what the building blocks should be.

An easy way to introduce these conjectures is to work on the similarity existing between the hierarchical model of aggregate resources defined in Chapter V and the well-known hierarchical organization advocated by Dijkstra (1969a) for the software of multiprogramming computer systems.

10.1 Levels of Abstraction

To quote Hoare: "The basic purpose of a computer operating system is to share the hardware which it controls among a number of users, making unpredictable demands upon its resources; and its objectives are to do so efficiently, reliably, and unobtrusively" (Hoare and Perrot, 1972, p. 11).

To this state description of an operating system may correspond the following process description.[1] The problem of constructing an operating system can be viewed as the problem of creating, usually by software, a set of "idealized" resources that control and allocate the hardware resources as efficiently, reliably, and conveniently as possible. These idealized resources should make "abstraction" of certain unavoidable hardware physical characteristics that are either too inefficient, unreliable, or difficult to live with (as are, for example, limitations in speed, quantity, or awkward and complicated access mechanisms).

There are two basic sources of complication in the problem of designing such a system:

(1) These idealized resources have to use or consume the very same hardware they control, and this is a source of conflict between efficiency and convenience.

(2) If these idealized resources are at all useful, the designer of the system should be their first beneficiary [because of his own "inability to do much" (Dijkstra, 1969b)]. Rather than using and being defined in terms of raw hardware resources, the operating system itself should use and be defined in terms of idealized resources. In other words, idealized resources should make use of each other, but the problem here is that this is a potential source of inefficiency and endless loops. It is also a design difficulty (where should the design be started, in which order?).

Dijkstra (1968, 1969a) showed that from a designer point of view it is advantageous to structure a computer operating system as a hierarchy of *levels of abstraction*. He considers an ordered sequence of machines $A_0, A_1, \ldots, A_l, \ldots$, where level A_0 is the given hardware and where the software of level l, $l = 0, 1, \ldots$, defined in terms of and executed by machine A_l, transforms machine A_l into machine A_{l+1}. The software of each level l creates an abstraction from some physical properties of the hardware. Examples of such abstractions implemented in the THE system (Dijkstra, 1968) are:

(1) At level 0, the central processor is allocated among concurrent processes so that the actual number of these processor(s) (one in this system) is no longer relevant: each process ready to use a processor may be considered above this level as having access to its own virtual processor;

(2) At level 1, the differences in mode and speed of access between a drum

[1] Simon (1962) sees "state description" and "process description" as the two main modes of understanding available to apprehend complex systems; briefly speaking, they correspond to the distinction between the world as sensed and the world as acted upon.

and a core memory are abstracted from in order to create a unique, homogeneous store. Above this level the physical location of information in this store is no longer relevant; information is identified by a segment name.

At the levels above, the unique operator console is transformed into a private one for each concurrent process, and input–output peripherals are transformed into more elegant logical devices.

A system like the THE system is thus built as a hierarchy of *abstract machines* $A_0, A_1, \ldots$. The role of an abstract machine is to "rebuild" a hardware resource into a more convenient "abstract resource," which will be used by the upper levels (Dijkstra, 1971). To do so, an abstract machine may only resort to the abstract resources created at the lower levels. This restriction is intended to facilitate the step-by-step construction and verification of the system. Such advantages are the direct consequence of the restricted nature of the interactions that are permitted among distinct levels. A more detailed discussion of this point would fall outside the scope of our work; interested readers may find information and other references in Dijkstra (1969a) and Parnas (1974).

However, this very same restriction that is imposed upon the interactions among distinct levels of the hierarchy leads to the following dilemma: if we decide to use the simplification introduced by abstracting from device R_l in producing the abstraction from device R_m, then we can no longer use the simplification introduced by abstracting from R_m in writing the programs that abstract from R_l. In a word, this restriction implies that each level helps only those above it. The choice of the ordering of the abstractions may therefore be quite difficult, and the point we would like to clarify here is to know whether the principles of near-complete decomposability that have been exploited throughout this monograph may help in this choice or not.

10.2 Aggregation and Ordering of Abstractions

It is possible to identify at least two types of design principles that must be obeyed to decide that the abstraction from a device R_i should be created at a lower level than the level of abstraction of a device R_j of another type.

(1) The abstraction from details of device R_i should be more *convenient* to program the abstraction from details of device R_j than conversely. If the abstraction from R_i is not more convenient, it need not come below the level of abstraction from R_j. This principle results directly from the restriction that an abstract machine may only use the abstract resources created at the lower levels.

(2) The use of an abstraction from R_i to program the abstraction from R_j should not prevent the latter resource from being controlled and allocated *efficiently*. This is obvious.

Condition (1) has the following consequence. We may expect that *in many cases* the degree to which an abstract resource is a "convenient" programming tool or concept will be reflected in the dynamics of the system by the "frequency" at which this abstract resource is accessed. An abstract resource seldom if ever used at execution time by another is likely to be a tool of little or no help in the design of this other. In this last proposition, the expression "used by another" must be interpreted in a broad sense; in an operating system, for example, the error handling routines are certainly convenient abstractions although they should normally be seldom invoked; similarly, as Parnas (1976) remarked, interrupt handling routines are convenient but are never explicitly called by other programs. These routines, however, do not violate our proposition if we take into account the fact that errors and interrupts demand a *constant* and vigilant detection. Thus, if we consider this detection (hardware or software) as part of error or interrupt handling, the abstract resources responsible for this handling appear to be in constant operation in the system; hence their convenience: they allow the designer to conceive the essential part of his system in an idealized, error- and interrupt-free universe, how frequent these events may be.

Taking these nuances into account, we may therefore expect that in general the convenience and the utility of an abstract resource will be reflected, when the system is in operation, by its frequency of utilization. In this case, the consequence of condition (1) is that the most heavily used abstract resources will be those implemented at the lower levels of the hierarchy.

The interpretation of condition (2) is more straightforward. Drums, disks, input devices, output devices, etc., are mutually asynchronous hardware devices capable of storing, retrieving, and transferring information at quite different speeds (see, e.g., Table 7.1). In order to control and allocate efficiently one of these devices, a program must be able to execute at a speed comparable with the speed of the device, i.e., the speeds of the resources (abstract or not) used by such a program should not be slower than the speed of the hardware it controls. This means that in order to operate efficiently an abstract machine should make use of resources, abstract or not, at least as fast as the hardware it is abstracting from. One could hardly imagine an abstract machine controlling the access to a drum and making use of magnetic tapes as working storage. The faster resources—fast being measured by the grain of time appropriate to describe the activity of the resource—must therefore be controlled and idealized at the lower levels of the hierarchy. This is indeed the case in the systems described in Dijkstra (1968) or Liskov (1972).

The application of conditions (1) and (2) to each type of abstraction leads therefore to an ordering such that the more frequently accessed abstract resources are implemented at the lower levels and also have faster access.

Since the speed of an abstract resource is geared by the speed of the hardware it controls, these two conditions lead inevitably to the situation where the fastest and more frequently used hardware is taken care of at the lowest levels of a hierarchy of abstractions; one can observe that this situation is indeed verified in most existing systems.

Since the lower its level the more frequently accessed and the faster the abstract resource of this level is likely to be, we conclude that the interactions between the levels of an "efficient and convenient" hierarchy of abstract machines are likely to obey the conditions for near-complete decomposability. This should not be a surprise: we had already reached the conclusion in Chapter VI that the conditions for a network to be balanced or optimal from the viewpoint of resource utilization conform to the conditions for near-complete decomposability.

We know also that the conditions for near-complete decomposability can be rather neatly defined in terms of parameters closely related to the usage and the performances of the interconnected resources of a system, that is, here, in terms of parameters related to the abstract resources and thus to the hardware that these abstract resources control Hence, provided these parameters can be evaluated, these conditions for near-complete decomposability could give the designer some guidance in the choice of the type of hardware that should be controlled at each level of abstraction and in the choice of the ordering of these levels so as to maximize the efficiency of the system and the convenience of each level of its hierarchical structure.

All this leads therefore to the fact that a level of abstraction as introduced above is likely to correspond to a level of aggregation. This correspondence can be carried further. We have seen that a level of aggregation may consist of several aggregates that can themselves be decomposed into subaggregates that belong to the level of aggregation below, and so on. Similarly, a level of abstraction is an abstract machine or a program that usually consists of several "modules," which can themselves be decomposed into "submodules" that belong to the level of abstraction below. Traditional criteria exist—not yet to mechanize the decomposition of a program into modules—but to characterize what a good or ideal module should be in order to meet the human constraints implied by team organizations and by the need of understandability. Holt (1975) mentions a few of these rules of thumb and observes that they all amount to stating that the complexity of a module's interactions with other modules should be less than the complexity of the module's internal structure, that is, modules should have simple interfaces, and the number of lines of interactions among interacting modules should, by far, be smaller than the lines of interaction within. If we substitute "line of interaction" by "degree of interaction" we come close to the definition of an aggregate. In so far as this substitution is valid, an aggregate is thus likely to correspond to a good module, and vice versa. A good functional decomposition of a system should be reflected by the dynamic behavior of this system.

10.3 Aggregation and Stepwise Design Evaluation

If a level of abstraction is likely to be assimilable to a level of aggregation, then we have the useful effect that a level of abstraction in an operating system is likely to be evaluable as a set of aggregates in a multilevel nearly completely decomposable system. In other words, the hierarchical organization of levels of abstraction that was originally presented to facilitate the step-by-step construction of the system turns out to facilitate its stepwise evaluation and performance prediction also.

Suppose, for example, that the model of the network of queues defined in Chapter IV is taken as a simplified model of a set of interacting abstract resources; the abstract machines $A_1, \ldots, A_l, \ldots$ could then be evaluated, respectively, by the aggregate resources $\mathcal{M}_1(n_1)$, $\mathcal{M}_2(n_2), \ldots, \mathcal{M}_l(n_l), \ldots$, each aggregate $\mathcal{M}_l(n_l)$ being a model of the equilibrium attained in the abstract machine A_{l+1} by the interactions between A_l and the abstract resource implemented at level l. By definition, the dynamic behavior of the aggregate $\mathcal{M}_l(n_l)$ toward its equilibrium may be evaluated with good approximation merely in terms of aggregative variables representative of the short-term equilibrium states attained at the lower levels of aggregation and without regard to the interactions with the upper levels. Since $A_1, A_2, \ldots$ represent as many distinct stages in the design, production, and testing of an operating system, such an evaluation technique that allows these stages to be evaluated independently should prove more valuable than techniques that may only be used after the design is completed.

As such, the hierarchy of aggregate resources studied in Chapter V is an example in the realm of analytical models of the level-by-level simulation techniques recommended by Parnas and Darringer (1967), Parnas (1969), Zurcher and Randell (1968), and Randell (1969). Of course, the formalism in which aggregate resources have been introduced in Chapter V does not map rigorously onto Dijkstra's abstract machines. Our aggregate model of Chapter V is but a simplified model of the levels of abstraction in an operating system, essentially because in our model a sequential process is waiting for or using at most one resource at a time, whereas in an operating system a sequential process may be requesting or using more than one resource simultaneously. But the ideas evoked here remain to a large extent independent of such differences.

To close this section, let us say that the above discussion reveals a property of these stages $A_1, A_2, \ldots$ of the design process: the rate of interaction between the components of a machine A_l, i.e., between A_{l-1} and the abstraction implemented at level $l-1$, is higher than the rate of interaction between the corresponding components of the next upper level A_{l+1}. Several arguments in favor of a decomposition of the design process into successive stages at which the subsystems coped with have a similar property, may be found elsewhere, in particular in Alexander (1964).

Conclusions

The discussion at the end of Chapter X revealed a property of the distinct stages $A_1, A_2, \ldots$ of the design process of a computer operating system: there is, on an average, more interaction between the system components dealt with at level A_l than there is between comparable components of the next upper level A_{l+1}. This property was tentatively grounded on exigencies of design (programming) convenience and system efficiency.

There is no doubt however that this property is but another facet of the more general proposition that states that complex systems are often hierarchic and that hierarchic systems are often nearly completely decomposable. This has been the recurrent theme of this work. It was fully exploited in an attempt to explain program locality and it was verified in all the systems we have investigated: queueing networks, memory hierarchies, user–machine interactions, operating systems. As such these systems, like the biological, social, economical, and physical systems mentioned by Simon (1962; Simon and Ando, 1961) bring empirical support for this proposition.

Simon was the first to make this proposition explicit and to set up a theoretical framework for it. Despite the depth and completeness of his work and of the work of his colleagues (Ando and Fisher, 1963; Fisher and Ando, 1962), many connections that probably exist between the concept of decomposability, certain properties of hierarchic structures, and the process of complex system design remain unclear or only empirically verified. More theoretical work is certainly needed to clarify these relations,[1] and in this respect the dichotomic distinction Simon (1962) himself makes between state description and process description as the two modes available to us for apprehending complex reality might very well be a starting point for such a research work.

[1] These relations have also been investigated by Alexander (1964), but from a slightly different point of view.

However, our purpose was less ambitious. Except on a few occasions—as in Chapter X, or when we showed in Chapter VI that balancing and optimization of resource utilization are factors contributing to near complete decomposability, and in Chapter VII that structure and efficiency are two qualities of a computer program that are not necessarily mutually exclusive—we did not try to support the claim that complex systems are naturally nearly completely decomposable. Instead we rather took it for granted and aimed at demonstrating its usefulness in such domains as queueing network and computer system performance analysis. The models studied in this work that in our opinion are the more successful in this respect are, on the one hand, those which show that the near-complete decomposability approximation can be an indispensable tool of analysis to cope within realistic computer time and space compass with the huge amount of calculations involved, and on the other hand, those which show this approximation as being vital in the understanding of certain stability phenomenons. In the demonstration of this usefulness, the finding of approximation bounds was obviously a key issue.

Various problems are raised and left unsolved by the present work. We have just mentioned the problem of the exact role played by hierarchic structures in the design process of complex systems, and the specific questions raised by Chapter IX have already been discussed in Section 9.15. In Chapter II, a procedure is given to obtain an estimate in ε^2 of the approximation made by aggregation; Theorem 2.2 suggests that an upper bound, not yet found, could exist for this estimate in terms of norms of the original and the aggregate matrices. In Chapter III, it is shown that necessary and sufficient conditions exist for a system to be nearly completely decomposable, and a criterion of sufficiency is proposed in terms of aggregate indecomposability and interdependency; a necessary condition expressed in similar terms might exist as well. The analysis of Chapter IX could be extended to open queueing networks with exogenous customer arrivals; different classes of customers requiring different types of service could also be considered as in Baskett *et al.* (1975).

A closing remark: We do not want to leave the reader with the false impression that we implicitly claim that near-complete decomposability is the more appropriate approach to solve the problems of computing systems performance evaluation and prediction. More powerful and adequate models will probably be found. Other techniques are already breaking through, such as those inspired from electrical circuit theory (Chandy *et al.*, 1975) or those based upon the diffusion approximation (Gaver and Shedler, 1971; Kobayashi, 1974; Gelenbe, 1975). Our purpose has been to propose one possible approach and to investigate its applicability as far as we could. Comparative studies of the pros and cons of the techniques as currently used will certainly be needed. In our opinion, however, in view of the complexity of the systems that have been and will have to be analyzed, the best chance of success will remain with decomposition methods that allow the analyst to exploit the old principle "divide and rule," and with approximation methods that allow him to trade in, in a controlled way, the luxury of accuracy for ease of analysis.

Another factor of success remains under the computer system designer's responsibility. "A machine is only useful provided the relevant aspects of its behavior are reasonably well predictable," as Dijkstra once wrote. When this principle becomes commonly accepted, and when sufficiently powerful models of performance prediction become available, any design that will not conform to these models will then have to be, as in any other branch of engineering, rejected.

APPENDIX I

Proof of Theorem 2.1

Relations (2.14) and (2.15) follow immediately from (2.5) and from the orthogonality of the left and right eigenvectors of **Q**. By setting $i = 1$ in Eq. (2.10) and isolating the elements $\beta_{1_I}(l_L)$, we have

$$(1-\lambda(l_L))\beta_{1_I}(l_L) + \varepsilon s^*(1_I)^{-1} \sum_{J=1}^{N} \beta_{i_J}(l_L)\gamma_{1_J 1_I}$$

$$= -\varepsilon s^*(1_I)^{-1} \sum_{J=1}^{N} \sum_{j=2}^{n(J)} \beta_{j_J}(l_L)\gamma_{j_J 1_I}, \qquad I = 1, \ldots, N, \qquad \text{(A1.1)}$$

which we regard as a linear nonhomogeneous system of N equations in N unknowns $\beta_{1_J}(l_L)$.

It follows from (2.9b) that

$$\gamma_{1_J 1_I} = n(I)^{-1} \sum_{j \in J} \sum_{i \in I} v^*_{j_J}(1_J) c_{j_J i_I},$$

and since for $J \neq I$, $c_{j_J i_I} = \varepsilon^{-1} q_{j_J i_I}$,

$$\gamma_{1_J 1_I} = (\varepsilon n(I))^{-1} p_{JI}, \qquad J \neq I. \qquad \text{(A1.2)}$$

We have also for any j_J that

$$\sum_{I=1}^{N} n(I)\gamma_{j_J 1_I} = \tilde{\mathbf{v}}^*(j_J)\left[\sum_{I=1}^{N} n(I)\,\mathbf{C}_{JI}\,\mathbf{v}^*(1_I)\right] = 0,$$

since the row sums of matrix **C** are zero and the elements of $\mathbf{v}^*(1_I)$ are all equal to $n(I)^{-1}$. Thus,

$$\gamma_{1_I 1_I} = -(\varepsilon n(I))^{-1} \sum_{J \neq 1} p_{IJ}. \qquad \text{(A1.3)}$$

Hence, if we introduce (A1.2) and (A1.3) into (A1.1), note that $s^*(1_I) = n(I)^{-1}$, and remember (Section 2.1.2) that all the $\beta_{j_J}(l_L)$, $j \neq 1$, on the right-hand side of (A1.1) are in ε if $l = 1$, we obtain the system (2.12), (2.13) stated in the theorem.

APPENDIX II

Proof of Theorem 2.2

For the kth element of the row vector $\tilde{\mathbf{B}}(I, L)[\lambda(1_L) I_{n(I)} - \mathbf{Q}_I^*]$ we obtain

$$a_k(I, L) = \lambda(1_L) \sum_{i=2}^{n(I)} \beta_{i_I}(1_L) v_{k_I}^*(i_I) - \sum_{l \in I} \sum_{i=2}^{n(I)} \beta_{i_I}(1_L) v_{l_I}^*(i_I) q_{l_I k_I}^*,$$

which is equal to

$$a_k(I, L) = \sum_{i=2}^{n(I)} \beta_{i_I}(1_L) v_{k_I}^*(i_I) (\lambda(1_L) - \lambda^*(i_I))$$

and, when $\beta_{i_I}(1_L)$, $i \neq 1$, is replaced by expression (2.11), becomes

$$a_k(I, L) = \varepsilon \sum_{J=1}^{N} \beta_{I_J}(1_L) \sum_{i=2}^{n(I)} s^*(i_I)^{-1} \gamma_{1_J i_I} v_{k_I}^*(i_I) + O(\varepsilon^2),$$

or

$$a_k(I, L) = \varepsilon \sum_{J=1}^{N} \beta_{1_J}(1_L) \sum_{j \in J} v_{j_J}^*(1_J) \sum_{l \in I} c_{j_J l_I}$$

$$\times \sum_{i=2}^{n(I)} s^*(i_I)^{-1} v_{l_I}^*(i_I) v_{k_I}^*(i_I) + O(\varepsilon^2),$$

since by definition

$$\gamma_{1_J i_I} = \sum_{l \in I} \sum_{j \in J} v_{j_J}^*(1_J) c_{j_J l_I} v_{l_I}^*(i_I).$$

Besides, by definition of matrices $\mathbf{H}^*$ and $\mathbf{H}^{*-1}$ (see Section 2.1.1), the summation

$$\sum_{i=1}^{n(I)} s^*(i_I)^{-1} v_{l_I}^*(i_I) v_{k_I}^*(i_I)$$

is the (l_I, k_I)th element of the product $\mathbf{H}^*\mathbf{H}^{*-1}$. Thus

$$\sum_{i=2}^{n(I)} s^*(i_I)^{-1} v^*_{m_I}(i_I)\, v^*_{k_I}(i_I) = \begin{cases} 1 - v^*_{k_L}(1_I), & \text{for} \quad m = k, \\ -v^*_{k_I}(1_I), & \text{for} \quad m \neq k. \end{cases}$$

Then we have

$$a_k(I, L) = \varepsilon \sum_{J=1}^{N} \beta_{1_J}(1_L) \sum_{j \in J} v^*_{j_J}(1_J)\left(c_{j_J k_I} - v^*_{k_I}(1_I)\right) \sum_{l \in I} c_{j_J l_I} + O(\varepsilon^2),$$

which proves the first part of Theorem 2.2. Moreover, for $L = 1$, by definition we have

$$\varepsilon c_{j_J l_I} = \begin{cases} q_{j_J l_I}, & \text{if} \quad J \neq I, \\ q_{j_I l_I} - q^*_{j_I l_I}, & \text{otherwise,} \end{cases}$$

and thus, by definition of the probabilities p_{IJ}, and by virtue of Theorem 2.1,

$$\begin{aligned} \varepsilon \sum_{J=1}^{N} \beta_{1_J}(1_1) \sum_{j \in J} v^*_{j_J}(1_J) \sum_{l \in I} c_{j_J l_I} \\ &= \sum_{J=1}^{N} \beta_{1_J}(1_1) \sum_{j \in J} v^*_{j_J}(1_J) \sum_{l \in I} q_{j_J l_I} \\ &\quad - \beta_{1_I}(1_1) \sum_{j \in J} v^*_{j_I}(1_I) \sum_{l \in I} q^*_{j_I l_I} \\ &= \sum_{J=1}^{N} \beta_{1_J}(1_1)\, p_{JI} - \beta_{1_I}(1_1)\, p_{II} = O(\varepsilon^2). \end{aligned}$$

Since the differences $\beta_{1_J}(1_1) - X_J(1)$ are themselves of order ε, we obtain

$$a_k(I, 1) = \varepsilon \sum_{j=1}^{N} X_J(1) \sum_{j \in J} v^*_{j_J}(1_J)\, c_{j_J k_I} + O(\varepsilon^2),$$

which completes the proof.

APPENDIX III

Numerical Example

A.3.1 Consider the matrix

$$\mathbf{Q} = \left[\begin{array}{ccc|cc|ccc}
0.85 & 0 & 0.149 & 0.0009 & 0 & 0.00005 & 0 & 0.00005 \\
0.1 & 0.65 & 0.249 & 0 & 0.0009 & 0.00005 & 0 & 0.00005 \\
0.1 & 0.8 & 0.0996 & 0.0003 & 0 & 0 & 0.0001 & 0 \\
\hline
0 & 0.0004 & 0 & 0.7 & 0.2995 & 0 & 0.0001 & 0 \\
0.0005 & 0 & 0.0004 & 0.399 & 0.6 & 0.0001 & 0 & 0 \\
\hline
0 & 0.0005 & 0 & 0 & 0.00005 & 0.6 & 0.2499 & 0.15 \\
0.00003 & 0 & 0.00003 & 0.00004 & 0 & 0.1 & 0.8 & 0.0999 \\
0 & 0.00005 & 0 & 0 & 0.00005 & 0.1999 & 0.25 & 0.55
\end{array}\right]$$

and the corresponding completely decomposable matrix

$$\mathbf{Q}^* = \left[\begin{array}{cccccccc}
0.85 & 0 & 0.15 & & & & & \\
0.1 & 0.65 & 0.25 & & & & & \\
0.1 & 0.8 & 0.1 & & & & & \\
 & & & 0.7 & 0.3 & & & \\
 & & & 0.4 & 0.6 & & & \\
 & & & & & 0.6 & 0.25 & 0.15 \\
 & & & & & 0.1 & 0.8 & 0.1 \\
 & & & & & 0.2 & 0.25 & 0.55
\end{array}\right] = \left[\begin{array}{ccc}
\mathbf{Q}_1^* & & \\
 & \mathbf{Q}_2^* & \\
 & & \mathbf{Q}_3^*
\end{array}\right].$$

The value of ε is 10^{-3} and matrix $\mathbf{C}$ is equal to

$$\left[\begin{array}{cccccccc}
0 & 0 & -1 & 0.9 & 0 & 0.05 & 0 & 0.05 \\
0 & 0 & -1 & 0 & 0.9 & 0.05 & 0 & 0.05 \\
0 & 0 & -0.4 & 0.3 & 0 & 0 & 0.1 & 0 \\
0 & 0.4 & 0 & 0 & -0.5 & 0 & 0.1 & 0 \\
0.5 & 0 & 0.4 & -1 & 0 & 0.1 & 0 & 0 \\
0 & 0.05 & 0 & 0 & 0.05 & 0 & -0.1 & 0 \\
0.03 & 0 & 0.03 & 0.04 & 0 & 0 & 0 & -0.1 \\
0 & 0.05 & 0 & 0 & 0.05 & -0.1 & 0 & 0
\end{array}\right].$$

The steady-state vectors and the eigenvalues of $\mathbf{Q}_1{}^*$, $\mathbf{Q}_2{}^*$, $\mathbf{Q}_3{}^*$ are, respectively,

$$\tilde{\mathbf{v}}^*(1_1) = [0.4 \quad 0.417391 \quad 0.182609],$$
$$\tilde{\mathbf{v}}^*(1_2) = [0.571429 \quad 0.428571],$$
$$\tilde{\mathbf{v}}^*(1_3) = [0.240741 \quad 0.555555 \quad 0.203704],$$
$$\lambda^*(1_1) = 1, \qquad \lambda^*(2_1) = 0.75, \qquad \lambda^*(3_1) = -0.15,$$
$$\lambda^*(1_2) = 1, \qquad \lambda^*(2_2) = 0.30,$$
$$\lambda^*(1_3) = 1, \qquad \lambda^*(2_3) = 0.55, \qquad \lambda^*(3_3) = 0.40.$$

The Bauer–Deutsch–Stoer upper bounds for the second largest eigenvalue (see Section 2.2.2) are equal, respectively, to 0.80 for $\mathbf{Q}_1{}^*$, to 0.30 for $\mathbf{Q}_2{}^*$, and to 0.55 for $\mathbf{Q}_3{}^*$.

The condition numbers of $\mathbf{Q}_1{}^*$, $\mathbf{Q}_2{}^*$, and $\mathbf{Q}_3{}^*$ are given by

$$s^*(1_1) = 0.333333, \qquad s^*(2_1) = 0.357142, \qquad s^*(3_1) = 0.448051,$$
$$s^*(1_2) = 0.5, \qquad s^*(2_2) = 0.5,$$
$$s^*(1_3) = 0.333333, \qquad s^*(2_3) = 0.321428, \qquad s^*(3_3) = 0.473684.$$

The aggregative matrix of the transition probabilities p_{IJ} between aggregates defined by (1.25) is

$$\mathbf{P} = \begin{bmatrix} 0.99911 & 0.00079 & 10^{-4} \\ 0.000614 & 0.999286 & 10^{-4} \\ 0.555 \times 10^{-4} & 0.445 \times 10^{-4} & 0.999 \end{bmatrix} \tag{A3.1}$$

and the steady state vector of $\mathbf{P}$ is

$$\tilde{\mathbf{X}}(1) = [0.222573 \quad 0.277427 \quad 0.5]. \tag{A3.2}$$

The approximation to the steady-state vector of $\mathbf{Q}$ yielded by the micro-variables $x_{i_I}(1) = X_I(1)\, v_{i_I}^*(1_I)$ is then

$$[x_{i_I}(1)] = [0.089029 \quad 0.0929 \quad 0.040644 \quad 0.15853$$
$$0.118897 \quad 0.12037 \quad 0.277777 \quad 0.101852].$$

In order to estimate the differences between $[x_{i_I}(1)]$ and the steady-state vector of $\mathbf{Q}$, first we must calculate the elements $a_i(I, 1)$ defined by (2.34); they can be obtained as the elements of the row vector $[x_{i_I}(1)] \times \varepsilon\mathbf{C}$:

$$[a_i(I, 1)] = 10^{-3}[0.067781 \quad 0.074523 \quad -0.142294 \quad -0.15466$$
$$0.015456 \quad 0.01080 \quad 0.007880 \quad -0.018681].$$

With these values one can solve the systems (2.32) and obtain the elements of the vectors $\mathbf{B}(I, 1)$:

$$[b_i(I, 1)] = 10^{-3}[0.271085 \quad -0.123778 \quad -0.147306 \quad -0.022080 \quad 0.022080 \quad 0.015082 \quad 0.017512 \quad -0.032594].$$

The elements of the vector $\mathbf{K}(\varepsilon, 1)$ can then be obtained by summing over each set I the elements of the row vector $[b_i(I, 1)] \times \mathbf{C}$:

$$\tilde{\mathbf{K}}(\varepsilon, 1) = 10^{-3}[-0.077169 \quad 0.077169 \quad 0].$$

By solving the 3×3 system defined by (2.18) we obtain the deviations of the macroprobabilities $X_I(1)$ from their correct values $\beta_{1_I}(1_I)$:

$$\beta_{1_1}(1_1) - X_1(1) = -0.051309 \times 10^{-3},$$
$$\beta_{1_2}(1_1) - X_2(1) = 0.051309 \times 10^{-3},$$
$$\beta_{1_3}(1_1) - X_3(1) = 0.$$

Using Eqs. (2.8), the differences $v_{i_I}(1_1) - x_{i_I}(1)$ may now be evaluated. Their values are given in Table A3.1 and compared with the exact differences.

The differences between the estimated and the exact values are in ε^2.

TABLE A3.1

	Estimated ($\times 10^{-3}$)	Exact ($\times 10^{-3}$)
$v_{1_1}(1_1) - x_{1_1}(1)$	0.250561	0.253397
$v_{2_1}(1_1) - x_{2_1}(1)$	−0.145194	−0.142388
$v_{3_1}(1_1) - x_{3_1}(1)$	−0.156676	−0.155546
$v_{1_2}(1_1) - x_{1_2}(1)$	0.007239	0.003436
$v_{2_2}(1_1) - x_{2_2}(1)$	0.044069	0.041099
$v_{1_3}(1_1) - x_{1_3}(1)$	0.015082	0.014981
$v_{2_3}(1_1) - x_{2_3}(1)$	0.017512	0.017753
$v_{3_3}(1_1) - x_{3_3}(1)$	−0.032594	−0.032733

A.3.2 The matrix $\mathbf{P}$ defined by Eq. (A3.1) is an example of a nearly completely decomposable matrix that is also row block stochastic:

$$\mathbf{P} = \mathbf{P}^{[1]} = \begin{bmatrix} \mathbf{P}_1^{*[1]} & \\ & \mathbf{P}_2^{*[1]} \end{bmatrix} + \varepsilon_2 \mathbf{C}^{[1]}$$

$$= \begin{bmatrix} 0.99911 & 0.89 \times 10^{-3} & 0 \\ 0.714 \times 10^{-3} & 0.999286 & 0 \\ 0 & 0 & 1 \end{bmatrix}$$

$$+ 10^{-4} \begin{bmatrix} 0 & -1 & 1 \\ -1 & 0 & 1 \\ 0.555 & 0.445 & -1 \end{bmatrix}.$$

One can verify that the steady-state vector of $\mathbf{P}_1^{*[1]}$ is [0.445137 0.554863], and the transition aggregative matrix between the aggregates $\mathbf{P}_1^{*[1]}$ and $\mathbf{P}_2^{*[1]}$ is

$$\mathbf{P}^{[2]} = \begin{bmatrix} 0.9999 & 10^{-4} \\ 10^{-4} & 0.9999 \end{bmatrix},$$

whose steady state vector is [0.5 0.5].

The resulting approximate steady-state vector of $\mathbf{P}$ is

$$x_{1_1}^{[1]} = 0.222569, \qquad x_{2_1}^{[1]} = 0.277431, \qquad x_{1_2}^{[1]} = 0.5.$$

The deviations from the vector $\mathbf{X}(1)$ [see Eq. (A3.2)] are thus very good; they are no more than 10^{-5}, while ε_2 is 10^{-4}.

Suppose now that $\mathbf{P}^{[1]}$ were not row block stochastic but equal to

$$\begin{bmatrix} 0.99911 & 0.88\times10^{-3} & 0.1\times10^{-4} \\ 0.614\times10^{-3} & 0.999286 & 10^{-4} \\ 0.555\times10^{-4} & 0.445\times10^{-4} & 0.9999 \end{bmatrix},$$

thus with weaker couplings between aggregates.

The steady-state vector of this matrix is

$$[0.267228 \quad 0.353025 \quad 0.379747].$$

Using the same aggregates as for $\mathbf{P}^{[1]}$, the aggregate transition matrix is

$$\begin{bmatrix} 0.99994 & 0.6\times10^{-4} \\ 10^{-4} & 0.9999 \end{bmatrix}$$

with steady-state vector [0.625391 0.374609]. The approximate steady-state vector is now [0.278385 0.347006 0.374609]; the maximal deviation is 0.111×10^{-1}, viz., of the same order of magnitude as

$$\frac{\varepsilon_2}{2(\eta_1+\varepsilon_2)} = \frac{10^{-4}}{2(0.88\times10^{-3}+10^{-4})} = 0.5\times10^{-1},$$

in accordance with our result of Section 2.5.

APPENDIX IV[1]

Eigenvalues of the Matrix $\mathbf{Q}(N, 1)$

The matrix $\mathbf{Q}(N, 1)$ of Lemma 4.2 has the form

$$\mathbf{M}_{n \times n} = \begin{bmatrix} 1-a & a & & & \\ b & 1-(a+b) & a & & \\ \cdots & \cdots & \cdots & \cdots & \\ & b & 1-(a+b) & a \\ & & b & 1-b \end{bmatrix},$$

where a and b are real positive numbers smaller than unity.

The eigenvalues of $\mathbf{M}$ are solutions of

$$P_n(\lambda) = \mathbf{det}(\mathbf{M} - \lambda \mathbf{I}_n) = 0.$$

If we set $y = (1-(a+b)-\lambda)$ in $(\mathbf{M} - \mathbf{I}\lambda_n)$, and replace each element of the last column by the corresponding row sum, we have

$$P_n(y) = (y+a+b)\,\mathbf{det} \begin{bmatrix} y+b & a & & & 1 \\ b & y & a & & 1 \\ \cdots & \cdots & \cdots & \cdots & \cdots \\ & b & y & a & 1 \\ & & b & y & 1 \\ & & & b & 1 \end{bmatrix};$$

If we carry successively on each row i, $i = 1, \ldots, n-1$, the operation

$$\text{row}(i) \leftarrow \text{row}(i) - \text{row}(i+1),$$

[1] We are indebted to Dr. Y. Genin, MBLE Research Laboratory, for these developments. The matrix $\mathbf{Q}(N, 1)$ is the Markov chain matrix of a random walk with reflecting barriers; a calculation of the eigenvalues of this type of matrix can also be found in Feller (1968, pp. 436–437).

and then on each column $j, j = 1, \ldots, n-2$, the operation

$$\mathrm{col}(j+1) \leftarrow \mathrm{col}(j+1) + \mathrm{col}(j),$$

we obtain

$$P_n(y) = (y+a+b)\,\mathbf{det}\begin{bmatrix} y & a & & & \\ b & y & a & & \\ \cdots & \cdots & \cdots & \cdots & \\ & b & y & a \\ & & b & y \end{bmatrix},$$

or

$$P_n(y) = (y+a+b)\,\Delta_{n-1}(y),$$

with

$$\Delta_0 = 1, \qquad \Delta_1 = y, \tag{A4.1}$$

and

$$\Delta_i = y\Delta_{i-1} - ab\Delta_{i-2}, \qquad i = 2, \ldots, n-1. \tag{A4.2}$$

The solution of this system of homogeneous difference equations is obtained as

$$\Delta_i = c_1 x_1{}^i + c_2 x_2{}^i, \tag{A4.3}$$

where x_1, x_2 are solutions of the characteristic polynomial (see, e.g., Ostrowski, 1960)

$$x^2 - yx + ab = 0.$$

From (A4.3) and (A4.1) it follows also that

$$c_1 + c_2 = \Delta_0 = 1,$$

$$c_1 - c_2 = \frac{y}{(y^2-4ab)^{1/2}}\,.$$

Then for $\Delta_{n-1}(y)$ we obtain

$$\Delta_{n-1}(y) = \frac{1}{2^n}\left[\frac{[y+(y^2-4ab)^{1/2}]^n}{(y^2-4ab)^{1/2}} - \frac{[y-(y^2-4ab)^{1/2}]^n}{(y^2-4ab)^{1/2}}\right].$$

Remembering that $y = -(a+b)$ yields the first eigenvalue $\lambda = 1$, the equation $\Delta_{n-1}(y) = 0$, of which the other eigenvalues are solutions, becomes

$$\left[\frac{y+(y^2-4ab)^{1/2}}{y-(y^2-4ab)^{1/2}}\right]^n = 1 = e^{j2K\pi},$$

which, after reduction, yields

$$y = \pm(4ab)^{1/2}\cos\frac{K\pi}{n}, \qquad K = 1, \ldots, n-1,$$

which is the result exploited in Lemma 4.2 of Chapter IV.

References

Aho, A. V., Denning, P. J., and Ullman, J. D. (1971). Principles of optimal page replacement, *J. Assoc. Comput. Mach.* **18**, 80–93.

Alexander, C. (1964). "Notes on the Synthesis of Form." Harvard Univ. Press, Cambridge, Massachusetts.

Ando, A., and Fisher, F. M. (1963). Near-decomposability, partition and aggregation, and the relevance of stability discussions, *Internat. Econom. Rev.* **4**, 53–67. *Reprinted in* Ando, A., Fisher, F. M., and Simon, H. A. (1963). "Essays on the Structure of Social Science Models." Harvard Univ. Press, Cambridge, Massachusetts.

Ara, K. (1959). The aggregation problem in input–output analysis, *Econometrica* **27**, 257–262.

Arora, S. R., and Gallo, A. (1971). The optimal organization of multiprogrammed multi-level memory, *Proc. ACM SIGOPS Workshop on System Performance Evaluation, Harvard Univ.*, pp. 104–141.

Avi-Itzhak, B., and Heyman, D. P. (1973). Approximate queueing models for multi-programming computer systems, *Operations Res.* **21**, 1212–1230.

Baskett, F., Chandy, K. M., Muntz, R. R., and Palacios, F. G. (1975). Open, closed, and mixed networks of queues with different classes of customers, *J. Assoc. Comput. Mach.* **22**, 248–260.

Bauer, F. L., Deutsch, E., and Stoer, J. (1969). Abschätzungen für die Eigenwerte positiver linearer Operatoren, *Linear Algebra and Appl.* **2**, 275–301.

Belady, L. A. (1966). A study of replacement algorithms for a virtual storage computer, *IBM Systems J.* **5**, 78–101.

Belady, L. A., and Kuehner, C. J. (1969). Dynamic space sharing in computer systems, *Comm. ACM* **12**, 282–288.

Belady, L. A., Nelson, R. A., and Shedler, G. S. (1969). An anomaly in space–time characteristics of certain programs running in a paging machine, *Comm. ACM* **12**, 349–353.

Betourne, C., and Krakowiak, S. (1972). Simulation de l'allocation de ressources dans un système conversationnel à mémoire virtuelle paginée, *Proc. Congrès AFCET, Grenoble*.

Birkhoff, G. (1948). "Lattice Theory." AMS Publ. **25**, Providence, Rhode Island.

Bodewig, E. (1956). "Matrix Calculus." North Holland Publ., Amsterdam.

Bryant, P. (1975). Predicting working set sizes, *IBM J. Res. Develop.* **19**, 221–229.

Burge, W. H., and Konheim, A. G. (1971). An accessing model, *J. Assoc. Comput. Mach.* **18**, 400–404.

Buzen, J. P. (1971a). "Queueing Networks Models of Multiprogramming." Ph.D. dissertation, Harvard Univ., Cambridge, Massachusetts.

Buzen, J. P. (1971b). Analysis of system bottlenecks using a queueing network model, *Proc. ACM SIGOPS Workshop on System Performance Evaluation, Harvard Univ.*, pp. 82–103.

Buzen, J. P. (1973). Computational algorithms for closed queueing networks with exponential servers, *Comm. ACM* **16**, 527–531.

Chandy, K. M., Herzog, U., and Woo, L. (1975). Parametric analysis of queueing networks, *IBM J. Res. Develop.* **19**, 36–42.

Coffman, E. G., Jr., and Ryan, T. A., Jr. (1972). A study of storage partitioning using a mathematical model of locality, *Comm. ACM* **15**, 185–190.

Coffman, E. G., Jr., and Wood, R. C. (1966). Interarrival statistics for time sharing systems, *Comm. ACM* **9**, 500–503.

Coolidge, J. L. (1959). "A Treatise on Algebraic Plane Curves." Dover, New York.

Courtois, P. J. (1972). On the near-complete decomposability of networks of queues and of stochastic models of multiprogramming computing systems, Sci. Rep. CMU–CS–72,111, Carnegie–Mellon Univ., Pittsburgh, Pennsylvania.

Courtois, P. J. (1975a). Decomposability, instabilities and saturation in multiprogramming systems, *Comm. ACM* **18**, 371–377. Copyright 1975, Association for Computing Machinery, Inc., adapted by permission.

Courtois, P. J. (1975b). Error analysis in nearly completely decomposable stochastic systems, *Econometrica* **43**, 691–709.

Courtois, P. J., and Georges, J. (1970). An evaluation of the stationary behavior of computations in multiprogramming computer systems, *Proc. ACM, Internat. Comput. Symp. Bonn, Germany* **1**, 174–182.

Courtois, P. J., and Georges, J. (1971). On a single-server finite queueing model with state-dependent arrival and service processes, *Operations Res.* **19**, 424–435.

Courtois, P. J., and Louchard, G. (1976). Approximation of eigencharacteristics in nearly completely decomposable stochastic systems, *Stochastic Processes and Their Applications.* **4**, 283–296.

Courtois, P. J., and Vantilborgh, H. (1975). Further results on instabilities and saturation in multiprogramming systems, MBLE Res. Rep. R317, Brussels.

Courtois, P. J., and Vantilborgh, H. (1976). A decomposable model of program paging behaviour, *Acta Informat.* **6**, 251–275.

Cox, D. R. (1962). "Renewal Theory." Methuen, London.

Cox, D. R., and Smith, W. L. (1954). On the superposition of renewal processes, *Biometrika* **41**, 91–99.

Cox, D. R., and Smith, W. L. (1961). "Queues." Methuen, London.

Denning, P. J. (1967). Effects of scheduling on file memory operations, *Proc. AFIPS 1967 SJCC* **30**, 9–21. AFIPS Press, Montvale, New Jersey.

Denning, P. J. (1968a). "Resource Allocation in Multiprocess Computer Systems." Ph.D. thesis, MAC-TR-50, MIT, Cambridge, Massachusetts.

Denning, P. J. (1968b). Thrashing: Its causes and prevention, *Proc. AFIPS 1968 FJCC* **33**, 915–922. AFIPS Press, Montvale, New Jersey.

Denning, P. J. (1968c). The working set model for program behavior, *Comm. ACM* **11**, 323–333.

Denning, P. J. (1970). Virtual memory, *Comput. Surveys* **2**, 153–189.

Denning, P. J. (1972). A note on paging drum efficiency, *Comput. Surveys* **4**, 1–3.

Denning, P. J., and Kahn, K. C. (1975). A study of program locality and lifetime functions, *ACM Operating Systems Rev.* **9**, 207–216.

Denning, P. J., and Schwartz, S. C. (1972). Properties of the working set model, *Comm. ACM* **15**, 191–198.

Dijkstra, E. W. (1968). The structure of the "THE" multiprogramming system, *Comm. ACM* **11**, 341–346.

Dijkstra, E. W. (1969a). Complexity controlled by hierarchical ordering of function and variability, *Software Eng.*, 181–185. Sci. Affairs Div., NATO, Brussels.

Dijkstra, E. W. (1969b). Notes on structured programming, Technische Hogeschool, Eindhoven. *Reprinted in* "Structured Programming," Dahl, O. J., Dijkstra, E. W., and Hoare, C. A. R. (1972), pp. 1–82. Academic Press, New York.

Dijkstra, E. W. (1971). Hierarchical ordering of sequential processes, *Acta Informat.* **1**, 115–138.

Durieu, J. (1974). "A Review of Sparse Matrix Solution Methods." MBLE Res. Rep. R250, Brussels.

Feller, W. (1968). "An Introduction to Probability Theory and Its Applications," Vol. I, 3rd ed. Wiley, New York.

Fine, G. H., Jackson, C. W., and McIsaac, P. V. (1966). Dynamic program behavior under paging, *Proc. ACM, Nat. Conf. 1966*, pp. 223–228. Thompson Book Co., Washington, D.C.

Fisher, F. M., and Ando, A. (1962). Two theorems on *ceteris paribus* in the analysis of dynamic systems, *Amer. Polit. Sc. Rev.* **56**, 1. *Reprinted in* Ando, A., Fisher, F. M., and Simon, H. A. (1963). "Essays on the Structure of Social Science Models," pp. 107–112. MIT Press, Cambridge, Massachusetts.

Franklin, M. A., and Gupta, R. K. (1974). Computation of page fault probability from program transition diagram, *Comm. ACM* **17**, 186–191.

Frazer, R. A., Duncan, W. J., and Collar, A. R. (1952). "Elementary Matrices." Cambridge Univ. Press, London.

Gaver, D. P., and Shedler, G. S. (1971). Multiprogramming system performance via diffusion approximations, IBM Res. Rep. RJ938 (#16259).

Gelenbe, E. (1971). The two-thirds rule for dynamic storage allocation under equilibrium, *Information Processing Lett.* **1**, 59–60.

Gelenbe, E. (1973). A unified approach to the evaluation of a class of replacement algorithms, *IEEE Trans. Comput.* **C-22**, 611–618.

Gelenbe, E. (1975). On approximate computer system models, *J. Assoc. Comput. Mach.* **22**, 261–269.

Gecsei, J., and Lukes, J. A. (1974). A model for the evaluation of storage hierarchies, *IBM Systems J.* **13**, 163–178.

Glowacki, C. (1974). Quelques résultats explicites sur certains algorithmes de remplacement de pages, Sci. Rep. MC/74/15, Univ. Catholic Louvain, Louvain.

Gordon, W. J., and Newell, G. F. (1967). Closed queueing systems with exponential servers, *Operations Res.* **15**, 254–265.

Graham, R. M. (1973). Performance prediction, *Lecture Notes Econom. Math. Systems* **81**, 395–463. Springer-Verlag, Berlin.

Green, H. A. J. (1964). "Aggregation in Economic Analysis." Princeton Univ. Press, Princeton, New Jersey.

Hatfield, D. J., and Gerald, J. (1971). Program restructuring for virtual memory, *IBM Systems J.* **10**, 168–192.

Haynsworth, E. V. (1959). Applications of a theorem on partitioned matrices, *J. Res. Nat. Bur. Standards* **62B**, 73–78.

Hoare, C. A. R., and McKeag, R. M. (1972). A survey of store management techniques, Pt. 2 *in* "Operating Systems Techniques" (C. A. R. Hoare and R. H. Perrot, eds.), pp. 132–151. Academic Press, New York.

Hoare, C. A. R., and Perrot, R. H. (1972). "Operating Systems Techniques." Academic Press, New York.

Holt, R. C. (1975). Structure of computer programs: A survey, *Proc. IEEE* **63**, 879–893.

Hopf, E. (1963). An inequality for positive linear integral operators, *J. Math. Mech.* **12**, 683–692.

Jackson, J. R. (1963). Jobshop-like queueing systems, *Management Sci.* **10**, 131–142.

Kaneko, T. (1974). Optimal task switching policy for a multilevel storage system, *IBM J. Res. Develop.* **18**, 310–315.

Kemeny, J. G., and Snell, J. L. (1960). "Finite Markov Chains." Van Nostrand-Reinhold, Princeton, New Jersey.

Khintchine, A. I. (1955). "Mathematical Methods in the Theory of Queueing." Griffin, London (1960 translation).

King, W., III (1971). Analysis of demand paging algorithms, *Proc. IFIP Congr. 1971* **1**, 485–490. North-Holland Publ., Amsterdam, 1972.

Kleinrock, L. (1968). Certain analytic results for time shared processors, *Proc. IFIP Congr. 1968* **2**, 838–845. North-Holland Publ., Amsterdam, 1969.

Knuth, D. E. (1968). "The Art of Computer Programming," Vol. I. Addison-Wesley, Reading, Massachusetts.

Kobayashi, H. (1974). Application of the diffusion approximation to queueing networks. Part I: Equilibrium queue distributions, *J. Assoc. Comput. Mach.* **21**, 316–328; Part II: Non equilibrium distributions and applications to computer modelling, *J. Assoc. Comput. Mach.* **21**, 459–469.

Koestler, A. (1967). "The Ghost in the Machine." Hutchinson, London.

Kral, J. (1968). One way of estimating frequencies of jumps in a program, *Comm. ACM* **11**, 475–480.

Liptay, J. S. (1968). Structural aspects of the system/360, model 85. The cache, *IBM Systems J.* **7**, 15–21.

Liskov, B. H. (1972). The design of the Venus operating system, *Comm. ACM* **15**, 144–150.

Little, J. D. C. (1961). A proof for the queueing formula: $L = \lambda W$, *Operations Res.* **9**, 383–387.

Lucas, H. C. (1971). Performance evaluation and monitoring, *Comput. Surveys* **3**, 79–91.

Lynn, M. S., and Timlake, W. P. (1969). Bounds for Perron eigenvectors and subdominant eigenvalues of positive matrices, *Linear Algebra* **2**, 143–152.

Madison, A., and Batson, A. (1976). Characteristics of program localities, *Comm. ACM* **19**, 285–294.

Marcus, M., and Minc, H. (1964). "A Survey of Matrix Theory and Matrix Inequalities." Allyn and Bacon, Boston, Massachusetts.

Mattson, R. L., Gescei, J., Slutz, D. R., and Traiger, I. W. (1970). Evaluation techniques for storage hierarchies, *IBM Systems J.* **9**, 78–117.

Mullery, A. P., and Driscoll, G. C. (1970). A processor allocation method for time-sharing, *Comm. ACM* **13**, 10–14.

Muntz, R. (1975). Analytic modelling of interactive systems, *Proc. IEEE* **63**, 946–953.

Muntz, R., and Baskett, F. (1972). Open, closed, and mixed networks of queues with different classes of customers, Tech. Rep. N33, Digital Systems Lab., Stanford Univ., Stanford, California.

Naur, P. (1965). The performance of a system for automatic segmentation of programs within an algol compiler (GIER ALGOL), *Comm. ACM* **8**, 671–676.

Ostrowski, A. M. (1960). "Solutions of Equations and Systems of Equations." Academic Press, New York.

Ostrowski, A. M. (1964). Positive matrices and functional analysis, *in* "Recent Advances in Matrix Theory," pp. 81–101. Univ. of Wisconsin Press, Madison, Wisconsin.

Parnas, D. L. (1969). More on simulation languages and design methodology for computer systems, *Proc. AFIPS 1969 SJCC* **34**, 739–744. AFIPS Press, Montvale, New Jersey.

Parnas, D. L. (1974). On a "buzzword": Hierarchial structure, *Proc. IFIP Congr. 1974* **2**, 336–339, North-Holland Publ., Amsterdam.

Parnas, D. L. (1976). Some Hypotheses about the 'Uses' Hierarchy for Operating Systems. Forschungsbericht BS I 76/1, Fachbereich Informatik, Technische Hochschule, Darmstadt.

Parnas, D. L., and Darringer, J. A. (1967). SODAS and a methodology for system design, *Proc. AFIPS 1967 FJCC* **31**, 449–474. AFIPS Press, Montvale, New Jersey.

Randell, B. (1969). Towards a methodology of computing system design, *Software Engineering*, 204–208. Sci. Affairs Div., NATO, Brussels.

Riordan, J. (1968). "Combinatorial Identities." Wiley, New York.

Rodriguez-Rosell, J. (1973). Empirical working set behavior, *Comm. ACM* **16**, 556–560.

Russell, B. (1948). "Human Knowledge, Its Scope and Limits." Allen & Unwin, London.

Schwartz, J. I., Coffman, E. G., and Weissman, C. (1964). A general purpose time-sharing system, *Proc. AFIPS 1964 SJCC* **25**, 397–411. Spartan Books, Baltimore, Maryland.

Schwartz, J. I., and Weissman, C. (1967). The SDC time-sharing system revisited, *Proc. 22nd Nat. Conf. ACM* **P-67**, 263–271. Thompson Book, Washington, D.C., and Academic Press, London.

Shedler, G. S., and Tung, C. (1972). Locality in page reference strings, *SIAM J. Comput.* **1**, 218–241.

Simon, H. A. (1962). The architecture of complexity, *Proc. Amer. Phil. Soc.* **106**, 467–482.

Simon, H. A. (1969). "The Sciences of the Artificial." MIT Press, Cambridge, Massachusetts.

Simon, H. A., and Ando, A. (1961). Aggregation of variables in dynamic systems, *Econometrica* **29**, 111–138. *Reprinted in* Ando, A., Fisher, F. M., and Simon, H. A., "Essays on the Structure of Social Science Models," pp. 64–91. MIT Press, Cambridge, Massachusetts, 1963.

Smith, J. L. (1966). An analysis of time sharing computer systems using Markov models, *Proc. AFIPS 1966 SJCC* **28**, 87–95. Spartan Books, Washington, D.C.

Smith, J. L. (1967). Multiprogramming under a page on demand strategy, *Comm. ACM* **10**, 636–646.

Spirn, J. R. (1973). "Program Locality and Dynamic Memory Management." Ph.D. thesis, Dept. Electr. Eng., Princeton Univ., Princeton, New Jersey.

Spirn, J. R., and Denning, P. J. (1972). Experiments with Program locality, *Proc. AFIPS 1972 FJCC* **41**, 611–621. AFIPS Press, Montvale, New Jersey.

Takács, L. (1960). "Stochastic Processes, Problems and Solutions." Methuen, London.

Takács, L. (1962). "Introduction to the Theory of Queues." Oxford Univ. Press, London and New York.

Taussky, O. (1948). Bounds for characteristic roots of matrices, *Duke Math. J.* **15**, 1043–1044.

Ten Hoopen, M., and Reuver, H. A. (1966). The superposition of random sequences of events, *Biometrika* **53**, 383–389.

Theil, H. (1957). Linear aggregation in input–output analysis, *Econometrica* **25**, 111–122.

Totschek, R. A. (1965). An empirical investigation into the behavior of the SDC time-sharing system, SDC Rep. SP-2131/000/00.

van Emden, M. H. (1969). Hierarchial decomposition of complexity, *Machine Intelligence* **5**, 361–382. Edinburgh Univ. Press, Edinburgh.

Vantilborgh, H. (1972). On random partially preloaded page replacement algorithms, MBLE Res. Rep. R202, Brussels.

Vantilborgh, H. (1974). On the working set size distribution and its normal approximation, *BIT* **14**, 240–251.

Varian, L., and Coffman, E. G. (1967). An empirical study of the behavior of programs in a paging environment, *Proc. ACM Symp. Operating System Principles 1st* [abstract in *Comm. ACM* **11**, 295–296].

von Bertalanffy, L. (1968). "General System Theory." Braziller, New York.

Wallace, V. L., and Mason, D. L. (1969). Degree of multiprogramming in page on demand systems, *Comm. ACM* **12**, 305–308.

Wallace, V. L., and Rosenberg, R. S. (1966). Markovian models and numerical analysis of computer system behavior, *Proc. AFIPS 1966 SJCC* **28**, 141–148. Spartan Books, Washington, D.C.

Weingarten, A. (1966). The Eschenbach drum scheme, *Comm. ACM* **9**, 509–512.

Wilkinson, J. H. (1965). "The Algebraic Eigenvalue Problem." Oxford Univ. Press (Clarendon), Oxford.

Zurcher, F. W., and Randell, B. (1968). Iterative multi-level modelling. A methodology for computer system design, *Proc. IFIP Congr. 1968* **2**, 867–871. North-Holland Publ., Amsterdam.

Index

A
B 7
C 8
D 9
E 0
F 1
G 2
H 3
I 4
J 5